Topic B1 — Cell Biology

Pages 2-3 — Cells

Warm-Up
Has a nucleus — Plant Cells and Animal Cells.
Contains a single loop of DNA — Bacterial Cells.
DNA is often found in plasmids — Bacterial Cells.
cytoplasm, cell membrane, cell wall

1 a) E.g.

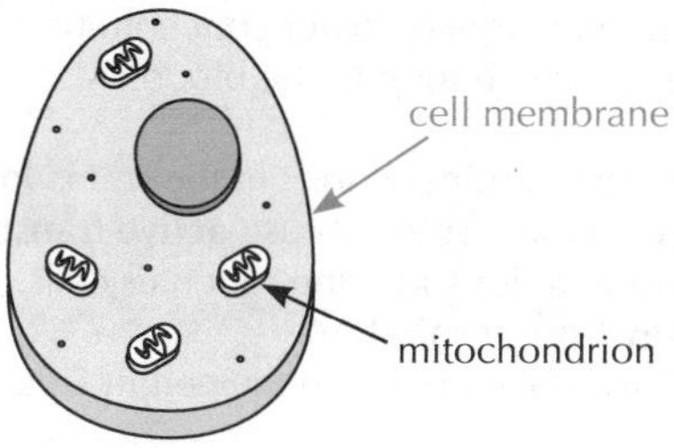

b) i) ribosomes
ii) E.g.

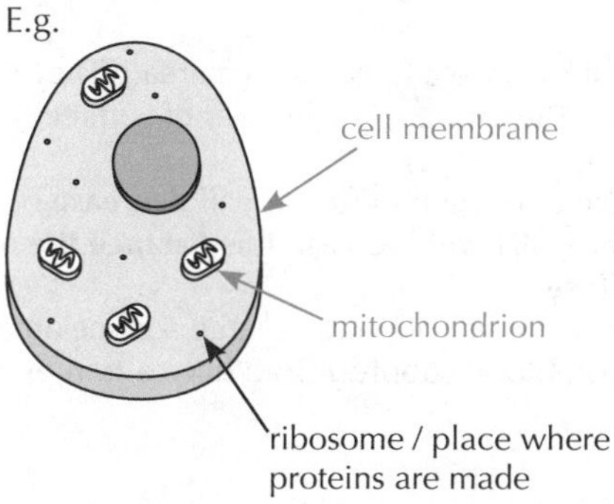

c) Any two from: e.g. it does not have chloroplasts. / It does not have a (permanent) vacuole. / It does not have a cell wall.

2 a) A
b) A
c) It does not have any mitochondria, and these are where most aerobic respiration happens.

3 a) i) chloroplasts
ii) inside the (permanent) vacuole
b) C

Pages 4-5 — Microscopy

Warm-Up
bigger, higher, more

1 D, E
2 To make the different parts of the cells easier to see.
3 Ribosomes are too small to be seen with a light microscope. They can only be seen with electron microscopes, which have a higher magnification and resolution.
4 C
5 a) i) magnification = 8 ÷ 0.02 = **× 400**
ii) 1 mm = 1000 µm
So 8 mm = 8 × 1000 = **8000 µm**
b) image size = 600 × 0.02 = **12 mm**

Page 6 — Cell Differentiation and Specialisation

Warm-Up
differentiation
E.g.

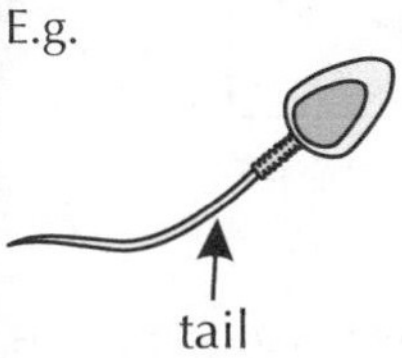

1 E.g.

Cell Type	[illegible]	[illegible]
	[illegible]	[illegible]
	Transports sugars around a plant.	Not many subcellular structures.
	Carries electrical signals around the body.	Long, with branches at the ends.
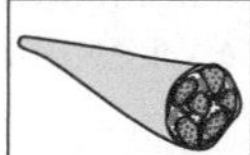	Contraction, which brings about movement.	Long, with many mitochondria.

From top to bottom, the table shows: xylem, phloem, a nerve cell and a muscle cell.

2 Mitochondria are where most of the reactions for aerobic respiration happen. Respiration transfers energy for the cell.

Pages 7-8 — Chromosomes and Mitosis

Warm-Up
DNA, pairs, double its DNA

1 A
2 A chicken has 78 ÷ 2 = 39 pairs of chromosomes.
A horse has 64 ÷ 2 = 32 pairs of chromosomes.
39 – 32 = **7 more pairs of chromosomes**
3 As an adult, to produce cells that are genetically different for reproduction.

Remember, mitosis produces genetically identical cells. Mitosis is used for growth and development in all multicellular organisms, but it's not used for reproduction in humans.

4 a) Before a cell divides, it copies (replicates) its DNA. X-shaped chromosomes form.
b) The chromosomes line up across the centre of the cell. The arms of each chromosome are pulled to opposite ends of the cell.
c) A membrane forms around each set of chromosomes. These become the nuclei of the new cells.
d) The cytoplasm and cell membrane divide. Two genetically identical cells are made.

5 a) Mitosis produces cells that are genetically identical to the original cells.
b) i) 20% of 24 = 24 × 0.2 = **4.8 hours**
ii) Any two from: e.g. the cells may be growing. / The cells may be increasing the amount of subcellular structures they have. / The cells may be replicating their DNA.

Pages 9-10 — Stem Cells

Warm-Up
Embryonic stem cell — Can divide to produce cells that can develop into <u>any type</u> of specialised cell.
Adult stem cell — Can divide to produce cells that can develop into <u>certain types</u> of specialised cells only.
E.g. bone marrow
E.g. blood cells

1 a) i) stem cells
ii) They are able to divide and differentiate into all the specialised cell types in the new plant.
b) red
c) E.g. the plant grown from the cutting will have identical characteristics to Norbert's original plant.

2 a) i) false
ii) true
iii) false
b) i) embryonic stem cells
ii) Embryonic stem cells can be made to turn into any type of specialised cell. Adult stem cells can only form certain ones (so may not be able to produce nerve cells).
iii) Embryonic stems cells come from embryos, and some people feel that these shouldn't be used, as each one is a potential human life.
c) If the stem cells are infected with a virus, they could pass it on to the patient receiving the stem cells and make them sick.

Pages 11-12 — Diffusion

Warm-Up
higher, lower
1 a) The squash particles have moved from an area of **higher** concentration (the **drop of squash**) to an area of **lower** concentration (the **water**).
b) slower
Diffusion is slower at lower temperatures because particles have less energy, so move around more slowly.
2 B, because the concentration gradient is steeper / there's a greater difference in concentration between the two sides.
3 a) E.g. the curry particles diffuse from an area of higher concentration (the kitchen) to an area of lower concentration (Madison's bedroom).
b) Adding more curry powder increases the concentration of the curry particles in the curry and so increases the rate of diffusion of the curry particles to the air.
c) Higher temperatures increase the rate of diffusion, so the curry particles will diffuse into the air faster when the curry is hot than when it is cold.
4 a) The glucose molecules diffused from an area of higher glucose concentration inside the bag to an area of lower glucose concentration outside the bag/in the water.
b) The starch molecules were too big to fit through the tiny holes in the membrane.
c) The longer, thinner bag has a greater surface area for the glucose to diffuse through, which sped up the rate of diffusion.

Pages 13-14 — Osmosis

Warm-Up
A membrane that only lets small molecules pass through it.
Water molecules move both ways across the membrane, but the overall movement is to the more concentrated solution.
1 a) into the cell
The solution outside the cell is less concentrated than the solution inside the cell.
b) A
2 a) i) Starting length = 50 mm
Final length = 53 mm
Change in length = **3 mm**
ii) Percentage change = (3 ÷ 50) × 100 = **6%**
b) 0.75 mol/dm^3
This doesn't fit the pattern — with the rest of the results, the percentage change has become more negative as the concentration of the solution increased.
c) E.g. 0.40 mol/dm^3 (accept answers > 0.25 mol/dm^3 and < 0.50 mol/dm^3)
3 Water molecules moved by osmosis from the pot (where they were in a higher concentration) into the ham (where they were in a lower concentration), so the ham increased in size.
This is just like in the famous osmosis in potatoes practical — when a potato cylinder is placed in a solution with a higher concentration of water than the potato cells, water moves into the potato cylinder, making it bigger.

Page 15 — Active Transport

Warm-Up

Feature	Diffusion	Active Transport
Substances move from areas of higher concentration to areas of lower concentration	✓	
Requires energy		✓

E.g. to move glucose from an area of lower concentration in the gut to an area of higher concentration in the blood.
1 a) root hair cell
b) The concentration of mineral ions in the soil is **lower** than in root hair cells. The cells use **active transport** to move the mineral ions into the root hair cells. This requires **energy** from **respiration**.
c) It has a smaller surface area than the cell in Figure 1.

Pages 16-17 — Exchanging Substances

Warm-Up
Any two from: e.g. they have a large surface area. / They have a good blood supply. / They have a single layer of surface cells.
1 a) B
b) E.g. the rate of oxygen diffusion will **decrease** because **the thicker walls will increase the distance the oxygen has to diffuse.**
c) Any two from: e.g. both have a large surface area. / Both have a good blood supply. / Both have a thin exchange surface.
2 a) i) smaller
ii) increases
iii) more
iv) slower
b) E.g. the rate of diffusion to cells deep in the body would be too slow.

Topic B2 — Organisation

Pages 18-19 — Cell Organisation

Warm-Up
cells, function, organ, organ system
1 a) i) Tissue
ii) E.g. it is a tissue because it is a group of cells with a similar structure and function.
b) E.g. Yes, Samina is correct. The different types of cells form different tissues. These work together to carry out a function (protecting the body).
2 a) i) The digestive system
ii) C
b) B
3 epithelial cell — 2, epithelial tissue — 3, heart — 4, circulatory system — 5

Pages 20-21 — Enzymes and Their Reactions

Warm-Up

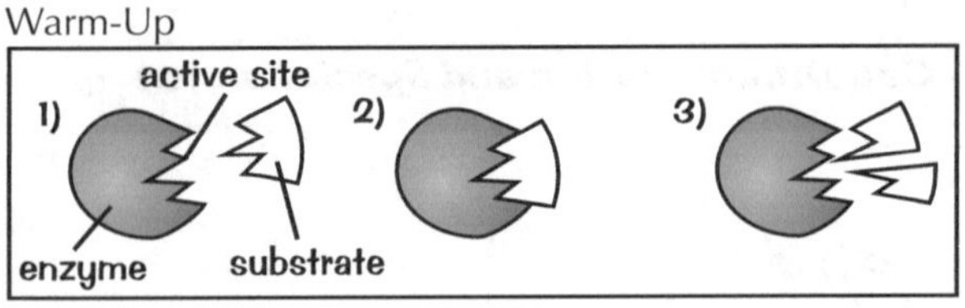

1 a) rate = amount of substance broken down ÷ time
= 10 ÷ 200
= **0.05 g/s**
b) E.g. the enzymes catalyse the breakdown of the fat that is blocking the sink.
c) i) E.g. it may have caused some of the bonds in the enzyme to break, changing the shape of the enzymes' active sites.

ii) E.g. the fat molecules won't fit into the enzymes' active sites any more. This means the enzymes can't catalyse the breakdown of the fat that is blocking the sink.

2 a)

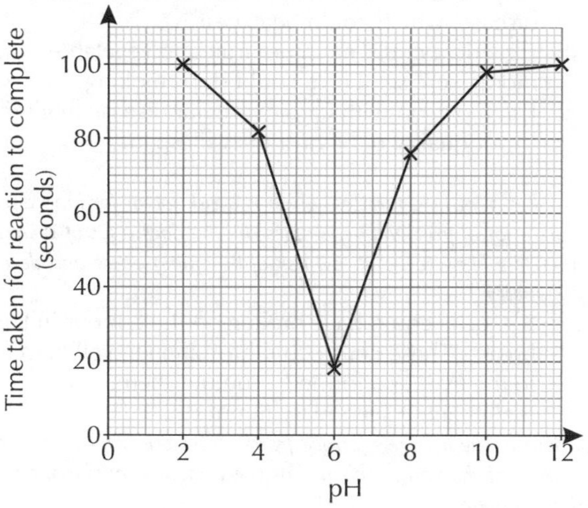

b) pH 6

c) No, because the enzyme doesn't work well at pH 2.

d) E.g. the temperature should have been the same at each pH / the volume of the substance and the enzyme should have been the same for each pH / the same method of deciding when the reaction is complete should have been used for each pH / the same measuring equipment should have been used for each pH.

Pages 22-23 — Enzymes, Digestion and Food Tests

Warm-Up

carbohydrates, starch, salivary glands

iodine test

1 a) Lipase(s)

b) B

The agar around species B has gone clear — this means that there aren't lipids in the agar. So species B must have lipase to break down the lipid.

2 a) Liver

b) B

Bile helps to neutralise hydrochloric acid from the stomach. If bile can't reach the small intestine because tube Y is blocked, then the small intestine will become more acidic. If this happens, the enzymes that break down fat in the small intestine won't work as well. The rate of fat breakdown will therefore decrease. Bile is also needed to emulsify fats, so if bile can't reach the small intestine, less fat will be emulsified.

3 a) proteins

b) E.g. the protease in the pineapple juice starts to break down/digest the collagen in the beef. This makes the beef softer and easier to chew.

4 a) Beaker **D** contains starch.
Beaker **C** contains a sugar solution.
Beaker **A** contains proteins.
Beaker **B** contains pure water.

b) Benedict's test

Pages 24-25 — The Heart and Lungs

Warm-Up

Alveoli — The small sacs in the lungs where gas exchange happens.

Bronchi — The tube-like structures that carry air from the trachea to the rest of the lungs.

Capillaries — Tiny blood vessels in the lungs.

1 a) A — pulmonary artery
B — vena cava
C — pulmonary vein

b) i) B

ii) the lungs

c) coronary arteries

2 a)

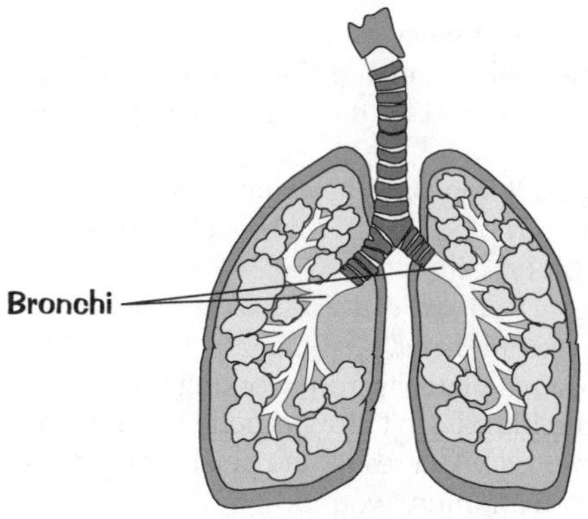

b) E.g.

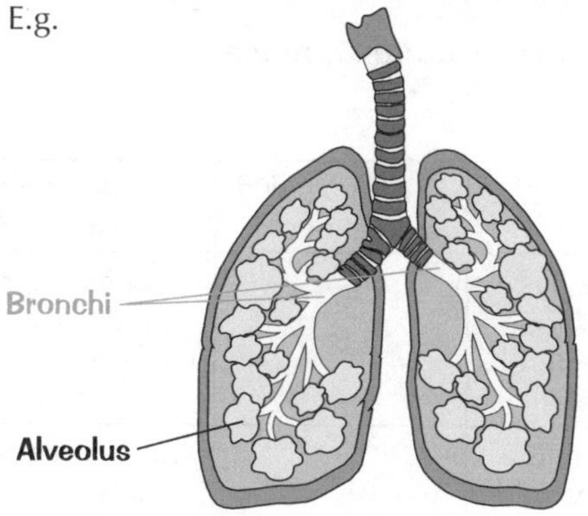

3 a) i) 240 seconds ÷ 60 = 4 minutes

$$\text{heart rate} = \frac{\text{number of heart beats}}{\text{number of minutes}} = \frac{140}{4} = \mathbf{35\ beats\ per\ minute}$$

ii) That Hugh has bradycardia.

b) E.g. Hugh could get an artificial pacemaker. This would keep Hugh's heart beating at a normal rate.

Pages 26-27 — Circulatory System — Blood Vessels

Warm-Up

away from, high

towards, low

thin/permeable, permeable/thin

1 a) B

b) i) arteries

Blood in the arteries is under higher pressure than blood in the capillaries and veins. The blood vessels represented by A have the highest blood pressure, so A must represent the arteries.

ii) The blood vessels represented by A have higher blood pressure. They therefore need more muscular walls so that they're strong enough to cope with the high blood pressure.

2 a) rate of blood flow = volume of blood ÷ time
= 0.9 ÷ 3
= **0.3 dm³/min**

b) 0.3 × 1000 = **300 cm³/min**

3 a) i) E.g. blood vessel B has a wider lumen than A has / blood vessel B has a thinner wall than A has.

ii) E.g. she could look to see whether it has valves.

b) Blood vessel A because it is the artery and arteries have elastic walls.

Pages 28-29 — Circulatory System — Blood

Warm-Up

Have a nucleus — white blood cells

Carry oxygen around the body — red blood cells

Help protect against infection — white blood cells

Shape gives them a large surface area — red blood cells

hormones, glucose, carbon dioxide

1 a) oxygen, lungs, tissues
b) More haemoglobin means that the whale can carry more oxygen in its blood, so it can go longer without breathing in.
2 a) E.g. it doesn't appear to have a nucleus, which is a feature of red blood cells. / It has a biconcave disc shape.
b) a white blood cell
When you're looking at a sample of blood under the microscope, any cell with a nucleus is a white blood cell.
3 E.g. BSS stops platelets from working properly. This means the blood can't clot properly when someone with BSS is injured. As a result, the person might lose lots of blood from an injury/wound.

Page 30 — Cardiovascular Disease

Warm-Up
fatty, reduce, oxygen
1 a)

	Recommended Treatment	Reason for recommendation
Annie has a high blood cholesterol level.	B	Statins will reduce the level of cholesterol in Annie's blood.
Valerie is a 40-year-old woman with heart failure. She may die soon if she is not treated.	A	Valerie needs a new heart soon, and may not have time to wait for a human heart transplant. An artificial heart will give her time.
Alastair has a blocked artery to the heart muscle.	D	Alastair needs a stent to open the artery to his heart muscle. (As his artery is already blocked, it's too late for him to use statins.)

b) E.g. there's a risk Annie could forget to take the statins. / Statins can cause unwanted side effects. / It can take time for the effects of statins to kick in.

Page 31 — Health and Disease

Warm-Up
Communicable diseases — Can be spread from person to person or between animals and people.
Non-communicable disease — Cannot be spread from person to person or between animals and people.
False
True
False
1 a) No — Anton may have some problems with his mental health, even if his physical health is OK.
b) Non-communicable
Communicable diseases spread between people because of disease-causing organisms (like bacteria and viruses). The faulty gene that causes cystic fibrosis can be passed from a parent to their child, but you can't 'catch' cystic fibrosis from another person like you can catch a cold.
c) E.g. an infection such as flu can trigger an immune system reaction. This could make the symptoms of Ollie's asthma worse.

Pages 32-33 — Risk Factors for Disease and Cancer

Warm-Up
smoking, sunbathing, drinking too much alcohol
1 a) 30%
b) E.g. the percentage of ex-smokers increased between 1974 and 2014.
c) C
d) E.g. the individual might not be able to work, which may mean they have less income than before. / The individual may have to move/adapt their home because of the disease, which could be expensive.
2 a) i) E.g. Tricia could eat more fruit and vegetables and do more exercise. Dave could stop smoking.
ii) E.g. a person with cardiovascular disease might have a lower quality of life/a shorter lifespan.
b) A
Dave smokes a lot of cigarettes, which is a risk factor for lung disease. Dave eats a lot of fruit and vegetables and he does a lot of exercise. He is therefore likely to be at low risk of becoming obese and developing Type 2 diabetes.
3 a) E.g. cells could have broken off the tumour in Freida's breast. These cells could have then travelled in the blood to the liver and formed a tumour there.
b) malignant, a secondary tumour
The tumour in the liver has formed from cancer cells that have spread from another part of the body. This means the tumour is malignant and it is a secondary tumour.
c) E.g. she might be worried that she has inherited a faulty gene that causes breast cancer from Freida.

Pages 34-35 — Plant Tissues, Transpiration & Translocation

Warm-Up
photosynthesis, spongy, transpiration, stomata
1 a)b)c)E.g.

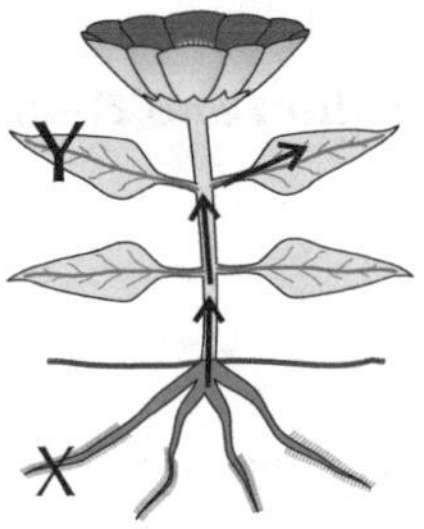

2 a) B
b) A
3 a) D
b) E.g. the amount of water in the vase will decrease. This is because water will be drawn up the carnation's xylem to replace the water lost through transpiration. The carnation flower will turn blue because water has travelled up the xylem to the flower, carrying blue dye with it.
4 a) xylem
Nitrate is a mineral ion. It's the xylem that transports mineral ions.
b) E.g. xylem tissue is made of (dead) cells joined end to end with no cell walls between them. It therefore forms a continuous tube that water and mineral ions can flow through. / The cell walls contain lignin, which makes the xylem stronger.

Pages 36-37 — Transpiration and Stomata

Warm-Up
size, fat, air
1 a) True
b) False
c) True
2 a) i) The humidity will increase.
Water moves from the leaves to the air through transpiration. Covering the plant with a clear plastic bag means that this water is trapped in the air surrounding the plant. The humidity of the air around the leaves will therefore increase.
ii) The humidity will decrease.
The water in the air surrounding the plant will get blown away by the fan. This means that the water from transpiration can't collect in the air around the leaves. The humidity of the air around the leaves will therefore decrease.

b)

Conditions	Line
not covered, not fanned	B
covered with clear plastic bag, not fanned	C
not covered, fanned	A

The higher the humidity, the lower the water loss from a plant. The humidity will be highest around the plant that is covered by a clear plastic bag. This plant will therefore have the least amount of water loss over time. The humidity will be lowest around the plant that is being fanned. This plant will therefore have the most water loss over time.

3 a) i) 0.2 – 0.1 = **0.1 cm**

ii) 3.6 – 1.4 = **2.2 cm**

ii) 4.5 – 4.2 = **0.3 cm**

b) i) C

The bubble moved further during the night than it did during the day. This means the cactus lost more water during the night. The results therefore suggest that most transpiration happens at night in cacti.

ii) The temperature is lower at night. A lower temperature means the rate of transpiration will be lower. Transpiring at night therefore means that the cactus will lose less water than if it transpired during the day.

Topic B3 — Infection and Response

Pages 38-39 — Communicable Disease

Warm-Up

Type of Pathogen: bacterium, virus, fungus, protist

NOT a Type of Pathogen: rhino, earthworm, mosquito, protest

E.g. in the air, in water, by direct contact.

1 a) the common cold — C, food poisoning caused by bacteria — C, thrush (an infection caused by a fungus) — C

b) cancer caused by smoking — N, Type 2 diabetes caused by obesity — N

2 a) E.g. because they could transfer the fungus that causes athlete's foot to the floor and other people could then catch athlete's foot/pick up the fungus by walking on the floor.

b) E.g. they should not share towels with other people.

3 a) The houses where people were infected are all downstream of the house where the first case happened.

b) C

The disease is not spread by vectors, so C won't work.

Pages 40-41 — Types of Communicable Disease

Warm-Up

protist, mosquitoes, vector

Gonorrhoea, HIV

E.g. measles

1 a) B

Only Salmonella is a pathogen that causes food poisoning. Gonorrhoea is a sexually transmitted disease caused by a bacterium, and TMV is a virus that affects plants.

b) C

c) E.g. wash his hands before and after preparing chicken / wash kitchen equipment used to handle raw chicken / keep chicken in the fridge.

2 a) Measles

b) Measles and HIV

c) HIV

d) Measles and HIV

e) Measles

f) HIV

3 a) A — rose black spot fungus

B — tobacco mosaic virus

b) i) E.g. they reduce the amount of photosynthesis carried out by the plant.

ii) If photosynthesis is reduced, less food/glucose will be produced, which plants need to grow.

4 E.g. water where mosquitoes lay eggs could be drained/ sprayed with insecticide.

Pages 42-43 — Fighting Disease

Warm-Up

They trap particles that could contain pathogens.

Surround a pathogen and digest it.

1 A bacterium that enters the body through the air and causes lung infections — cilia and mucus in the bronchi.

A bacterium that enters the body in food — stomach acid.

A fungus that enters the body through cuts — the skin and the substances it produces.

2 a) It starts to increase.

b)

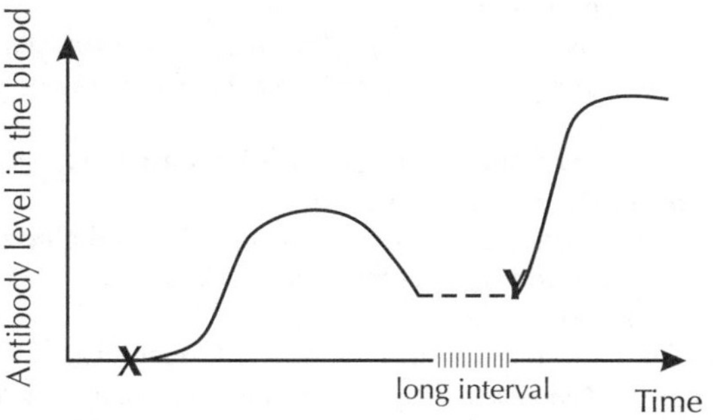

3 a) i) 4, 1, 3, 2

ii) The chickenpox virus does not produce toxins (only bacteria produce toxins).

b) John's white blood cells are able to rapidly produce antibodies against the virus when he is infected for a second time. The virus is quickly destroyed and he doesn't get ill.

Pages 44-45 — Fighting Disease — Vaccination

Warm-Up

True, True, False, False

1 white blood cells, inactive, antigens, specific

2 a) 5000

b) C

c) E.g. yes because there was a big reduction in the number of polio cases from 1956 onwards.

d) There are still a few cases of polio across the world, so the virus hasn't been completely wiped-out. So people are still vaccinated in case they come into contact with someone in/from another country who carries the virus.

Pages 46-47 — Drugs and Developing Drugs

Warm-Up

painkiller

aspirin — willow, digitalis — foxgloves, penicillin — mould

1. Human cells in a laboratory
2. Mammals (other than humans)
3. Healthy human beings
4. Patients with the disease

1 a) C

b) D

c) double-blind trial

d) E.g. yes, it is very effective at reducing 'bad' cholesterol, as it achieves a much greater reduction in this than the placebo (52% compared to 7%).

2 a) The medicine doesn't kill the virus causing the cold — it just relieves the symptoms.

b) Colds are caused by viruses and antibiotics only kill bacteria.

c) Because viruses reproduce using your own body cells, which makes it very difficult to develop drugs that destroy just the virus without killing the body's cells.

Topic B4 — Bioenergetics

Page 48 — Photosynthesis

Warm-Up
chloroplasts, water, oxygen, endothermic
D

1 a) insoluble, doesn't cause
b) Any two from: e.g. for respiration / for making cellulose/ cell walls / for making amino acids.
c) Through photosynthesis.

When the plants grow above ground, they can use light to produce glucose in photosynthesis.

Pages 49-50 — The Rate of Photosynthesis

Warm-Up
faster, too little, chlorophyll

1 a) E.g. temperature and light intensity.

The temperature and the amount of light during the day will be different between summer and winter, so the rate of photosynthesis will be different.

b) A

2 a) The rate of photosynthesis will **increase** because **the light intensity has increased**.
b) The temperature increases, which causes the enzymes involved in photosynthesis to work faster.
c) i) decrease
ii) E.g. the rate of photosynthesis will decrease.

The herbs use carbon dioxide for photosynthesis, so if there are a lot of herbs in the room the carbon dioxide concentration will decrease. This will then cause the rate of photosynthesis to decrease.

3 a) i) 40 units
ii) 33 °C (accept answers between 31 °C and 35 °C)
iii) 45 °C

When the enzymes are denatured, photosynthesis won't happen at all. At 45 °C, the rate of photosynthesis is 0 units.

b) i) E.g. the volume of oxygen produced (in a set time).
ii) E.g. he could have used a gas syringe to collect the oxygen produced. He could have then measured the length/volume of the oxygen bubble that formed.

Pages 51-52 — Respiration and Metabolism

Warm-Up
glucose, oxygen, water
$C_6H_{12}O_6$
C, D

1 a) C
b) E.g. metabolic reactions are controlled by enzymes. Enzymes are denatured at high temperatures.

2 a) i) E.g. it helps them to keep warm.
ii) E.g. to build up larger molecules from smaller ones and to move about.
b) i) less energy
ii) E.g. they wouldn't transfer enough energy to keep all their metabolic reactions going.
iii) glucose → lactic acid

3 a) i) carbon dioxide
ii) water
b) E.g. respiration is an exothermic reaction that transfers energy to the environment. This energy increases the temperature of the test tube.
c) E.g. she could repeat the experiment using a test tube containing dead/boiled seeds (as a control).

Page 53 — Exercise

Warm-Up
E.g. heart rate/breathing rate/breath volume increases.

1 a) At the end of the race Jim's breathing rate is 45 breaths per minute.
45 – 15 = **30 breaths per minute**
b) C
c) E.g. his breath volume will increase to get more oxygen into his blood and therefore into his cells.
d) E.g. Jim starts respiring anaerobically and lactic acid builds up in his muscles. This causes muscle fatigue/ tiredness.

Topic B5 — Homeostasis and Response

Page 54 — Homeostasis

Warm-Up
C

1 a) 3, 2, 4, 1
b) i) high blood pressure
ii) heart
c) E.g. the brain receives and processes information and then organises a response.
d) A

Pages 55-56 — The Nervous System, Synapses and Reflexes

Warm-Up

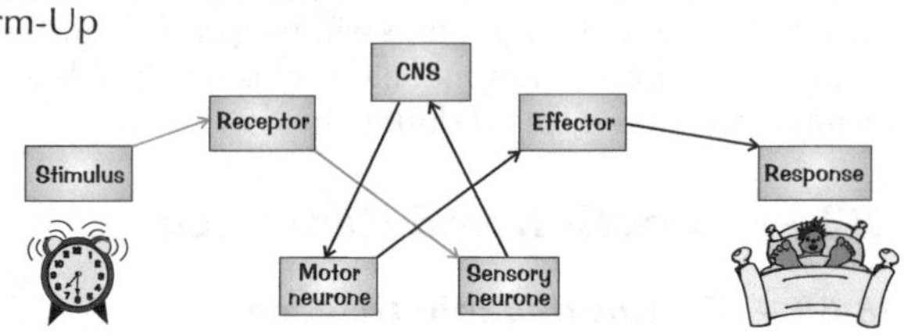

The brain and spinal cord.

1

	A	B
stimulus	pressure of the pin	smell of the food
receptor	pressure receptors	smell receptors
effector	muscles in the leg	salivary glands
response	foot is lifted up	saliva released into mouth

2 a) Jamie dropped the lasagne immediately.
Jamie dropped the lasagne without thinking about it.
b) Reflex reactions can help to stop you from getting injured.

3 a) i) sensory neurone
ii) motor neurone

Remember, sensory neurones carry information from receptors to the CNS. Motor neurones carry information from the CNS to effectors.

b) relay neurone
c) effector
d) i) synapse
ii) The nerve signal is transferred across the synapse/ junction X by chemicals, which move across the gap.

Page 57 — The Endocrine System

Warm-Up

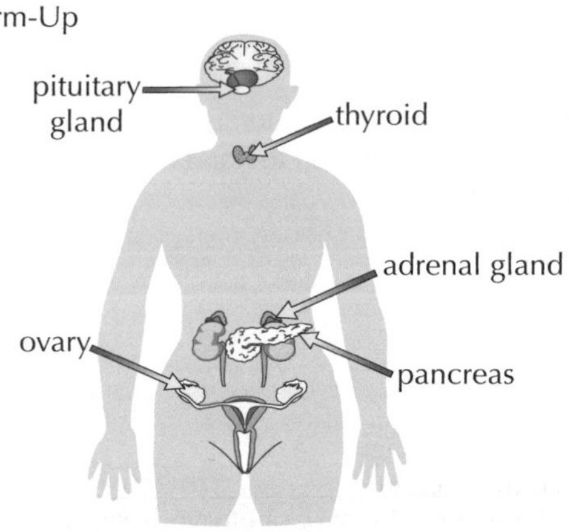

1 a) nervous system
b) hormonal system

c) nervous system
d) hormonal system
2 E.g. the hormones produced by the 'master gland' act on other glands, which can affect the release of hormones from these glands. So if the 'master gland' isn't working properly, many systems in the body may be affected.

Pages 58-59 — Controlling Blood Glucose

Warm-Up
pancreas, pancreas, glycogen
B, D
1 a) i) E.g. glucose from the digested meal is absorbed into Ruby's blood.
ii) E.g. the insulin causes glucose to move from Ruby's blood into her liver/muscle cells.
b) i) B
When you exercise, a lot more glucose is removed from the blood.
ii) less insulin
The amount of insulin a person with diabetes needs will vary depending on things like how much carbohydrate they eat and how much exercise they do. As exercise removes additional glucose from the blood, it may reduce the amount of insulin Ruby needs to take.
2 a) i) 16 380 ÷ 31 500 = 0.52
0.52 × 100 = **52%**
ii) 31 500 × (95/100) = **29 925 people**
b) E.g. adults may be more likely to be obese/very overweight, which increases the likelihood of developing Type 2 diabetes.
3 a) Type **2** diabetes because **the body is not responding to the insulin produced.**
b) E.g. getting regular exercise.

Pages 60-61 — Puberty and the Menstrual Cycle

Warm-Up

Role	FSH	LH	Progesterone	Oestrogen
Causes an egg to mature in an ovary	✔			
Causes the release of an egg		✔		
Involved in the growth and maintenance of the uterus lining			✔	✔

1 a) testosterone
b) testes
c) E.g. it causes facial hair to grow. / It stimulates sperm production.
2 An egg matures in the ovaries but ovulation does not take place. — LH
An egg does not mature in the ovaries. — FSH
Bleeding occurs between periods. — progesterone
LH causes the release of an egg (ovulation). FSH causes an egg to mature in an ovary. Progesterone helps to maintain the uterus lining. If the level of progesterone drops, the lining can break down and bleeding can occur even if a period is not due.
3 The average menstrual cycle is 28 days.
35 – 28 = 7 days
7 ÷ 28 = 0.25 × 100 = **25%**
4 oestrogen
5 a) 2, 4, 1, 3
b) i) 3 / An egg is released after maturing in the ovary.
ii) E.g. LH causes the release of an egg from the ovary.
c) E.g. progesterone maintains the uterus lining, so its level needs to drop for the lining to break down.

Page 62 — Controlling Fertility

Warm-Up
Contraceptive injection — Hormonal
Diaphragm — Non-hormonal
Abstinence — Non-hormonal
Contraceptive patch — Hormonal
Condom — Non-hormonal
1 a) E.g. stop fertilised eggs from implanting in the uterus wall / it prevents maturation/release of eggs.
b) spermicide
2 a) E.g. the contraceptive patch/implant/injection / intrauterine device.
The contraceptive patch is effective for a week at a time. Implants can last for three years and injections last two to three months. Intrauterine devices also last several years.
b) E.g. sterilisation is a permanent procedure.

Topic B6 — Inheritance, Variation and Evolution

Page 63 — DNA

Warm-Up
chromosomes, double helix, genes
1 gene, chromosome, nucleus, cell
2 a) E.g. scientists have studied and worked out the entire human genome.
b) E.g. it could help scientists to develop treatments for breast cancer / understand breast cancer better.

Pages 64-65 — Reproduction and Meiosis

Warm-Up
fusion, meiosis, half the
1 b) The parent cell makes copies of its DNA. This makes X-shaped chromosomes.
c) The parent cell splits in two. Each new cell gets half of the chromosomes.
d) Each new cell divides. There are now four gametes. Each gamete has half the original number of chromosomes.
2 a) Asexual reproduction
b) Sexual reproduction
c) Sexual reproduction
d) Asexual reproduction
e) Sexual reproduction
f) Sexual reproduction
g) Asexual reproduction
h) Asexual reproduction
3 a) i) C
Gametes contain half the number of chromosomes of a normal cell: 40 ÷ 2 = 20.
ii) 40 chromosomes
You could have worked out the answer here using a calculation (20 + 20 = 40) or you might have recognised that the fertilised egg cell would have the same number of chromosomes as a body cell.
b) i) mitosis
ii) E.g. the cells differentiate / the cells become more specialised.

Page 66 — X and Y Chromosomes

1 Person **B** is male because **they have an X and a Y chromosome.**
2 a)

	Z	W
Z	ZZ	ZW
Z	ZZ	ZW

b) 50% (or 0.5 or 1 in 2)

Pages 67-68 — Genetic Diagrams

Warm-Up
E.g. fur colour in mice / red-green colour blindness in humans.
Dominant allele — The allele whose characteristic is shown even if only one copy is present. — capital letter, e.g. 'A'
Recessive allele — The allele whose characteristic is only shown if two copies are present. — small letter, e.g. 'a'

1 a) i) The rabbit has **agouti** fur colour because **the allele for agouti fur colour (F) is dominant.**

ii) heterozygous

b) i) C and D

ii) A

The allele for solid fur colour is recessive, so both its alleles need to be recessive for a rabbit to show this characteristic.

2 a) QQ and Qq

b) i)

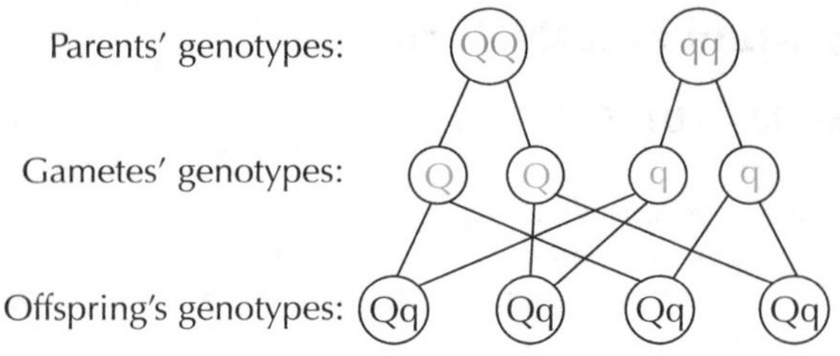

ii) A

c) i)

parent's alleles

parent's alleles	Q	q
Q	QQ	Qq
q	Qq	qq

ii)

parent's alleles

parent's alleles	Q	q
Q	QQ	Qq
q	Qq	qq

Pages 69-70 — Inherited Disorders and Family Trees

Warm-Up

E.g. screening suggests that people with these problems are not wanted and may lead to them being treated unfairly. / Screening may lead to embryos being destroyed/pregnancies being terminated. / People might want to start screening their embryos so they can pick the features they prefer. / Screening is expensive.

1 a) E.g. yes because polydactyly can cause babies to be born with extra fingers.

b) E.g. the person will **have polydactyly** because **the disorder is caused by a dominant allele and the person has the dominant allele.**

c) i)

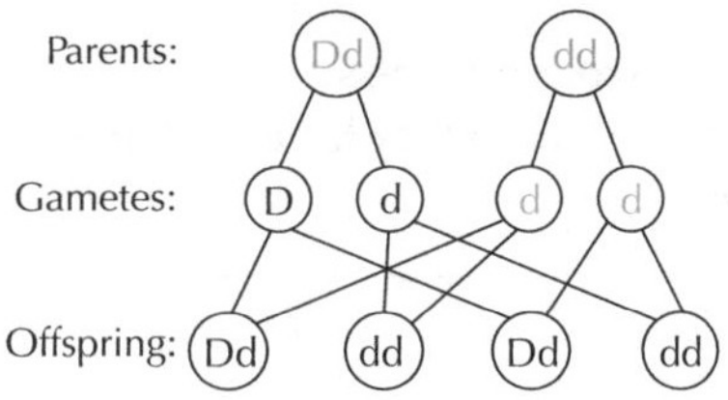

ii) B

Two of the four offspring have the dominant allele for polydactyly and so will have the disorder. 2/4 is the same as 0.5.

2 a) Any two from: e.g. Oliver / Mirna / Angus.

b) E.g. plenty of the family carry the allele but don't have the disorder.

c) i) B

ii)

Paul's alleles

Eilidh's alleles	F	f
F	FF	Ff
f	Ff	ff

Pages 71-72 — Variation and Evolution

Warm-Up

random, continuously, very rarely

C

1 A catastrophic event kills every member of the species — A rare plant species only lives on one volcano. They are all wiped out when the volcano erupts.

The environment changes too quickly — An island's rainforest is completely chopped down. All of the striped monkey's habitat is destroyed.

A new disease kills every member of the species — A new fungal pathogen is accidentally introduced to a habitat. It kills every member of a species of toad.

2 a) Environment

b) Genes

c) Genes

d) Environment

e) Both

3 a) B

Variation in racing ability is likely to be due to a combination of genes and the environment. Cinnamon was unlikely to become a champion racer if he was not raised in an environment that made him good at racing.

b) genes

4 2 — Birds were less likely to spot and eat the moths that blended in with their environment. Birds were more likely to eat the moths that stood out like a sore thumb.

3 — So the stick-like moths were more likely to survive and reproduce.

4 — Genes that made the moths look like sticks were more likely to be passed on to their offspring.

Page 73 — Antibiotic-Resistant Bacteria

Warm-Up

Doctors should only prescribe antibiotics when they really need to.

Patients should take all the antibiotics a doctor prescribes for them.

1 a) E.g. they were killed by the antibiotic. / They were outcompeted by the antibiotic-resistant bacteria.

b) E.g. the number of (antibiotic-resistant) bacteria increased. / There was an increase in the proportion of bacteria that were antibiotic-resistant.

c) C

Pages 74-75 — Selective Breeding and Genetic Engineering

Warm-Up

D

genome, characteristic, insulin

1 a) Any two from: e.g. the modern carrot is a darker/more orange than the wild carrot. / The modern carrot is larger than the wild carrot. / The modern carrot is less branched than the wild carrot.

b) E.g. so that it produces more food / is nicer to look at.

c) E.g. the breeder could select the carrot plants that are most resistant to the disease/least badly affected by the disease and breed them together. The best of these offspring/the most resistant offspring could be selected and bred together again. This process could be repeated over several generations. Eventually, all offspring will have a high resistance to leaf blight.

2 a) A

b) E.g. maize crop A, because it has the highest yield and making a crop herbicide-resistant should help to increase its yield / genetically modifying a crop usually leads to increased yields.

c) i) E.g. if weeds have the genes for herbicide resistance, farmers won't be able to use herbicide to kill the weeds.

ii) E.g. some people are worried we might not understand the effects of GM crops on human health.

Page 76 — Fossils

Warm-Up

in rock

1 a) E.g. it was too cold for decay microbes to work.

b) A

2 a) X

New layers of rock form on top of old layers, so more recent fossils are usually found in layers higher up in the rock.

b) E.g. animals tend to have more hard body parts than plants (e.g. bones or shells). These parts are fossilised more easily than soft body parts.

Pages 77-78 — Classification

Warm-Up

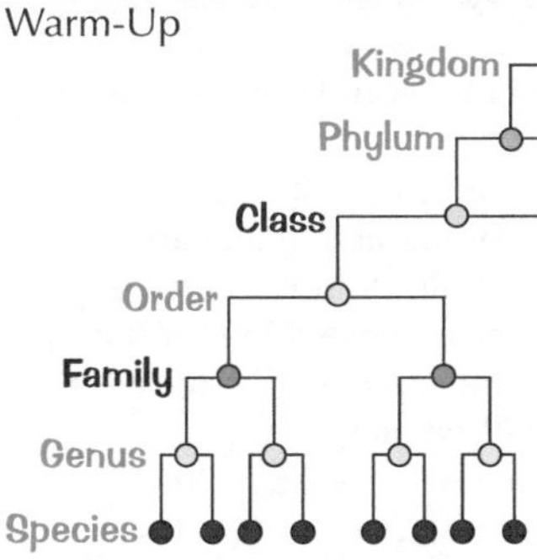

Eukaryota, Bacteria, Archaea

1 a) *Somateria*

b) *crecca*

c) Mallard and Teal

d) Eukaryota

2 a) True

X is further away from dolphins, mice, rays and sharks than Z on the evolutionary tree. This means X lived longer ago than Z.

b) False

c) True

d) True

3 a) i) D

ii) C

iii) D

Species D is the only one with the same genus name as species A. So species D is the most closely related to species A.

iv) C

Species A, B and D are all in the same family. Species C is in a different family, so it's the most distantly related to species A.

b)

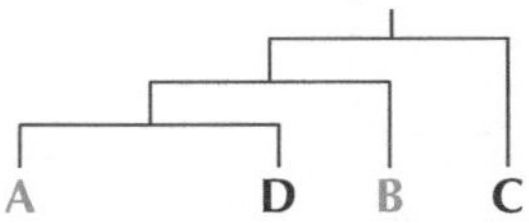

Topic B7 — Ecology

Pages 79-80 — Competition, Abiotic and Biotic Factors

Warm-Up

territory, food, mates

Any two from: e.g. new predators / competition / new diseases/pathogens / food availability.

1 a) C

The snakes might have been outcompeting the lizards for territory. This could have caused the number of lizards to decrease. The lizard population would probably increase if they were eating snakes. Organisms only compete with members of their own species for mates, so the lizards won't have been competing with the snakes for mates.

b) i) E.g. there were fewer lizards/predators to eat the crickets.

ii) B

Both weevils and crickets eat willow. The increase in the number of crickets meant there was more competition for willow.

2 a) April

b)

Factor	
water temperature	✓
higher number of predator fish	
more diseases of algae	
lower number of predator fish	✓
light intensity	✓

3 a) i) Water inputs = 38 300 + 700 = **39 000 m³**

ii) Water outputs = 34 800 + 4200 = **39 000 m³**

b) Yes, because the inputs of water to the ecosystem balance the outputs of water from the ecosystem.

Pages 81-82 — Adaptations

Warm-Up

Behavioural adaptation — A way that an organism is adapted to act.

Functional adaptation — An adaptation related to things that go on inside an organism's body.

Structural adaptation — An adaptation that is a feature of an organism's body parts.

1 B — Functional adaptation

C — Behavioural adaptation

D — Structural adaptation

2 a) E.g. it has a long tail / long ears / large feet.

b) A

c) E.g. the shape of the long-eared jerboa means that it can lose heat more easily and can therefore survive better in the hot environment where it lives.

3 a) i) Light intensity will be lower around deep sea vents than in shallow coastal waters.

ii) The water pressure will be higher around deep sea vents than in shallow coastal waters.

b) There isn't enough light in the deep sea for photosynthesis.

c) Species A, because it's adapted to live in an extreme environment that has high temperatures/high pressure/ no light.

d) E.g. Species A has adaptations that allows it to survive the conditions around deep sea vents. Species B does not have these adaptations.

Pages 83-84 — Food Chains

Warm-Up

Prey — An animal that is hunted and eaten by other animals.

Producer — An organism found at the start of a food chain.

Predator — An animal that hunts and eats other animals.

1 a) sea lettuce

b) starfish

c) B

The periwinkle eats other organisms so it is a consumer. The periwinkle doesn't eat animals so it is not a predator.

d) C

2 a) Predator: heron

Prey: frog

b) B

c) a stable community

The numbers of predators and prey are cycling, which is what happens in a stable community.

3 a) oak tree, photosynthesis, biomass, caterpillar

b) E.g. I agree, because the oak tree produces biomass using energy from the Sun. Without the oak tree, there would be no biomass/energy available to be passed on to the other organisms in the food chain. OR
I don't agree, because other organisms could be considered as important in the food chain as the oak tree. For example, the caterpillar passes biomass/energy along to all the other consumers in the food chain.

Pages 85-86 — Using Quadrats and Transects

Warm-Up

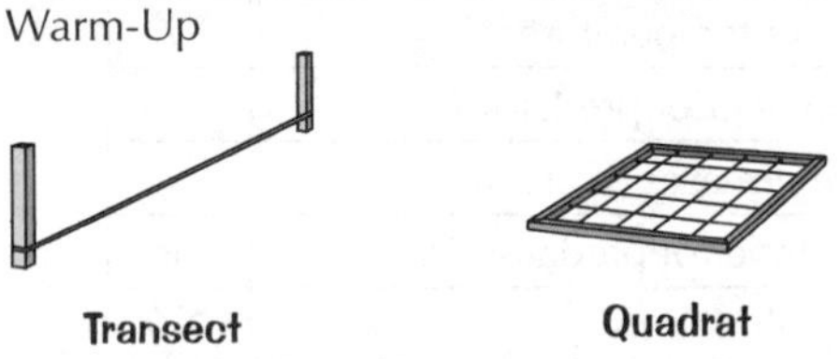

A, D

1 a) B

b) 2 — Count all the harebells within the quadrat.
3 — Repeat the previous steps at several other random points.
4 — Divide the total number of harebells by the total number of quadrats to find the mean.

2 a) The results in order are 0, 1, 2, 3, 4.
So the median is **2 gnomes per quadrat**.

b) i) Mean = $\frac{\text{total number of gnomes}}{\text{number of quadrats}}$

$= \frac{(3+1+2+4+0)}{5}$

= **2 gnomes per quadrat**

ii) The quadrat is 0.5 m^2.
$1 \div 0.5 = 2$
$2 \times 2 =$ **4 gnomes per 1 m^2**

c) $160 \times 4 =$ **640 gnomes**

3 a) E.g.

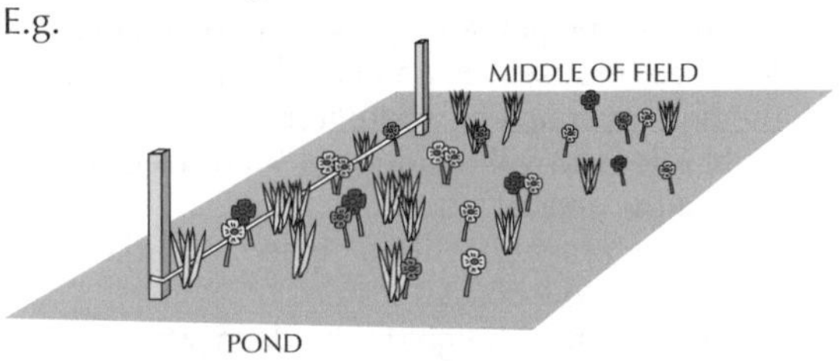

b) E.g. it means Sandy has a larger sample size for her data. This helps to make sure her results are valid.

Pages 87-88 — The Water Cycle and The Carbon Cycle

Warm-Up
Evaporation — A
Precipitation — C
carbon dioxide

1 a) evaporates, plants, condenses

b) i) It must cool down.

ii) E.g. air is pushed upwards when it hits the mountain. This causes the water vapour in the air to cool and condense. This means lots of clouds form above the mountain.

c) E.g. the water cycle provides trees with fresh water for photosynthesis.

2 a), b) and c)

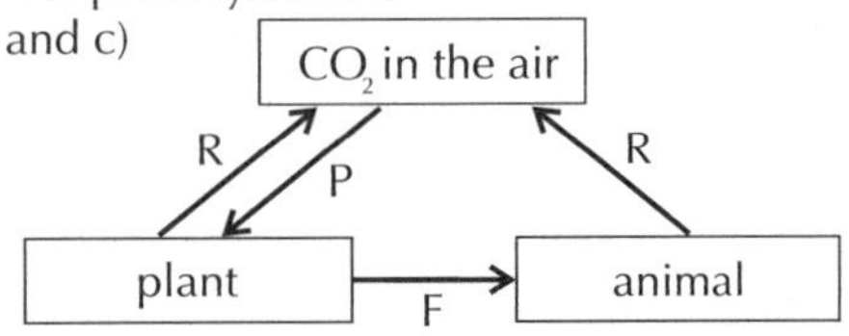

3 a) 2 — Plants take in carbon dioxide for photosynthesis and make carbon compounds.
3 — Animals eat the plants' carbon compounds.
4 — Plants and animals die.

b) E.g. dead organisms are broken down by microorganisms, which release carbon dioxide during respiration. / Burning of wood/fossil fuels/animal products releases carbon dioxide into the air.

Pages 89-90 — Biodiversity, Waste and Global Warming

Warm-Up
Smoke — Air
Household waste sent to landfill — Land
Sewage — Rivers and streams
E.g. methane.

1 a) John buys more belongings, which use more resources to manufacture. John drives a car and Derek rides a bicycle.

Buying more belongings means more energy is used to make them and more waste is produced disposing of them. A car also uses more energy and produces more waste than a bicycle. Recycling reduces waste, so that's not the reason that John's ecological footprint is bigger.

b) Any sensible suggestion, e.g. John could buy fewer new things / use his car less.

c) E.g. people in developed countries have access to more luxuries that make life more comfortable (e.g. cars, computers, etc.) than people in developing countries. More energy will be used to make these things and they will produce more waste, so the ecological footprint of the person who buys them will be bigger.

2 a) i) Total number of species in lake A: 15 + 75 = 90
Total number of species in lake B: 2 + 4 = 6
90 – 6 = **84 more species**

ii) Lake A, because it has a greater number/variety of fish and other species than lake B.

b) i) The perch will have less food to eat.

ii) Those in lake B.

The perch in lake A will still be able to feed on sticklebacks and roach. So the number of perch in lake A may not decrease much. The perch in lake B are completely dependent on minnows for food. So there is likely to be a large decrease in the number of perch in lake B.

iii) E.g. a perch population is more likely to be stable when there is greater biodiversity.

In an ecosystem with greater biodiversity, the perch are less dependent on any one species for survival. This means that the population of perch is less likely to change over time.

c) E.g. the temperature of the lake could become too high for some species to survive / rainfall patterns could change, causing the lake to dry out.

Pages 91-92 — Land Use and Maintaining Biodiversity

Warm-Up
True, False, False, True

1 a) B

The species of monkey is rare and at risk of dying out. It is therefore called an 'endangered' species. A 'common' species is one that's found in large numbers or lots of places. An 'extinct' species has no individuals left alive.

b) E.g. to provide land for farming/cattle/rice fields/ growing crops for biofuels.

c) E.g. other species that live in the forest will be protected / the biodiversity of the forest will be protected / it could help to slow global warming.

d) E.g. monkeys could be taken into captivity and a breeding programme could be set up.

e) E.g. the scientists want the monkey's habitat to be replanted.

2 a) i) E.g. removing peat from a bog destroys the habitat. This means fewer animals, plants and microorganisms will be able to live there.

ii) E.g. less food waste needs to be sent to landfill, so less land will be needed for landfill. This means more land may be available as a habitat.

b) C

Peat bogs need to be drained so the peat can be used in compost. When the peat is drained, microorganisms can break it down. They release carbon dioxide when they respire. This contributes to global warming.

3 a) ... forests are often cut down to make way for large fields that contain a single biofuel crop.

b) Fields containing one crop often have a very low biodiversity. Allowing grasses and wildflowers to grow around the crop could increase biodiversity by providing a habitat for more species.

Topic C1 — Atomic Structure and the Periodic Table

Pages 93-94 — Atoms and Elements

Warm-Up

Atoms always have a charge of **zero**. An atom which has lost or gained electrons is called an **ion**. A neutral atom has the same number of electrons and **protons**. If an electron is added to a neutral atom, the atom becomes **negatively** charged.

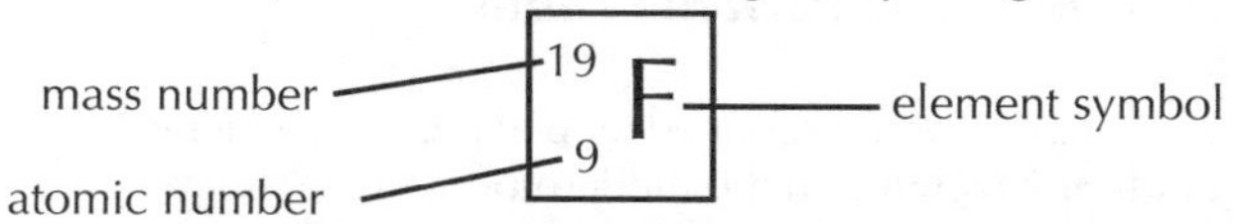

1

Element	Symbol	Mass Number	Number of Protons	Number of Electrons	Number of Neutrons
Sodium	Na	**23**	11	**11**	**12**
Neon	**Ne**	**20**	10	10	10
Calcium	Ca	**40**	**20**	20	20

2 a) the nucleus

b) E.g.

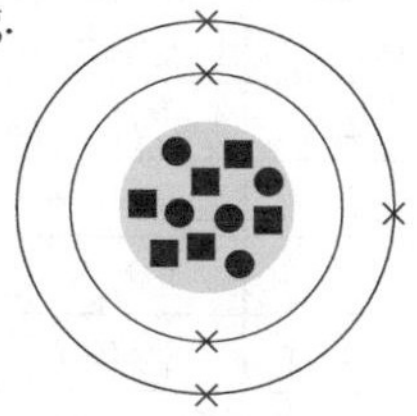

There should be 5 circles and 6 squares, all in the nucleus.

c) boron

3 a) $^{12}_{6}\mathrm{C}$ $^{14}_{6}\mathrm{C}$

b) These two atoms have the same atomic number / number of protons but a different mass number / number of neutrons.

4 $A_r = \dfrac{\text{sum of (isotope abundance} \times \text{isotope mass number)}}{\text{sum of abundances of all the isotopes}}$

% isotope abundance of Ga-71 = 100 – 60 = 40

$A_r = \dfrac{(60 \times 69) + (40 \times 71)}{100}$

$= \dfrac{4140 + 2840}{100}$

= 69.8

= **70** (to 2 s.f.)

Pages 95-96 — Compounds and Chemical Equations

Warm-Up

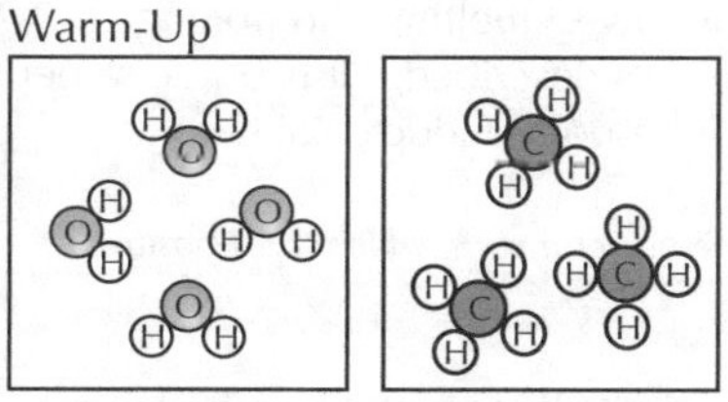

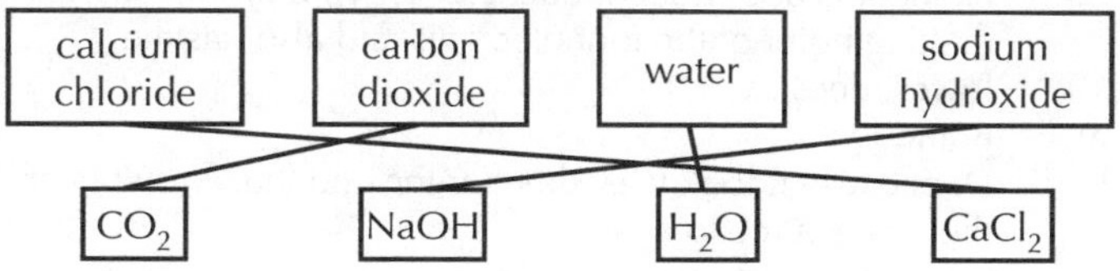

1

Compound	Formula	Number of elements	Number of hydrogen atoms
Hydrochloric acid	HCl	**2**	**1**
Methane	CH_4	**2**	**4**
Calcium hydroxide	$Ca(OH)_2$	**3**	**2**
Ethanol	C_2H_5OH	**3**	**6**

2 a) i) potassium, bromine

ii) potassium bromide

b) $2K + Br_2 \rightarrow 2KBr$

3 a) $N_2 + 3H_2 \rightarrow \mathbf{2}NH_3$

b) $\mathbf{4}Fe + \mathbf{3}O_2 \rightarrow 2Fe_2O_3$

c) $4NH_3 + \mathbf{5}O_2 \rightarrow 4NO + 6H_2O$

4 a) i) lithium + oxygen → lithium oxide

ii) Li_2O

iii) Compounds contain atoms of more than one element. Oxygen molecules are made up of oxygen atoms only, and so oxygen is an element / is not a compound.

b) $\mathbf{2}Li + \mathbf{2}H_2O \rightarrow \mathbf{2}LiOH + H_2$

Pages 97-98 — Mixtures and Chromatography

Warm-Up

A

1 a) The brown, green and orange sweets.

b) Different dyes travel up the paper at different rates and separate out. For the brown, green and orange sweets, two or more spots of dye formed. This shows that they contain two or more dyes that have separated out.

2 C

The chemical properties of the copper and magnesium are not changed by mixing them together. So the magnesium will still react with the acid, and the copper will still not react.

3 Any two from: e.g.

Mistake — the spots of dye/ink are touching the solvent. Correction — the student should put the filter paper in a beaker of solvent with the pencil line above the level of the solvent.

Mistake — the ink and dyes are compared using different solvents. Correction — the student should use the same solvent for the black ink and the dyes so that it's a fair test.

Mistake — the ink and dyes are compared on different pieces of filter paper which could make it difficult to directly compare them. Correction — the student should put the spots of the dyes and the ink on the same piece of filter paper.

Pages 99-100 — More Separation Techniques

Warm-Up

C

1 a) B and D

b) D

c) E.g. the boiling points of diethyl ether and ethyl ethanoate are far enough apart for them to be separated by simple distillation, as well as by fractional distillation. The boiling points of ethyl ethanoate and THF are too similar for simple distillation to be used. So they can only be separated by fractional distillation.

2 Mei should remove the evaporating dish from the heat and leave the solution to cool. When crystals have formed, she should filter them out of the solution and leave them in a warm place to dry.

3 a) i) Contents of the flask: propanol

Contents of the beaker: methanol

ii) Methanol has a lower boiling point than propanol and so will evaporate off first and be collected in the beaker.

b) Jo heated the mixture above the boiling points of both methanol and propanol. So both liquids evaporated from the flask and were collected in the beaker.

Pages 101-102 — The History of the Atom

Warm-Up

James Chadwick

1

Discovery	Model of the atom that was used until the discovery	Model of the atom that was developed using the discovery
The discovery of the electron	**Atoms are solid spheres**	**The plum pudding model**
When alpha particles are fired at a thin sheet of gold, some are deflected more than expected. Some are even deflected backwards.	**The plum pudding model**	**The nuclear model**

2 a) Mistake 1: Dylan has indicated that the electrons are positively charged with the '+' signs.
Correction 1: Dylan should replace the '+' signs with '–' signs.
Mistake 2: Dylan has labelled the 'pudding' as neutral.
Correction 2: Dylan should change the label to say the 'pudding' is positively charged.

b)

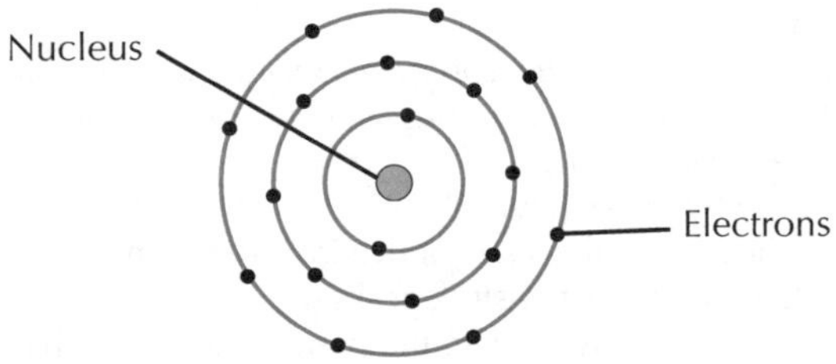

Pages 103-104 — Electronic Structure

Warm-Up

A and C

Look up the atomic number of the element (in the periodic table). This tells you the number of protons in atoms of that element, which is equal to the number of electrons.

1 a) 2, 8, 8, 1

b) E.g.

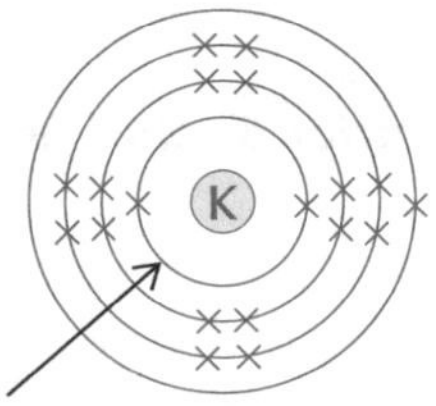

It doesn't matter where you've drawn the arrow, as long as it points to the electron shell closest to the nucleus.

c) E.g.

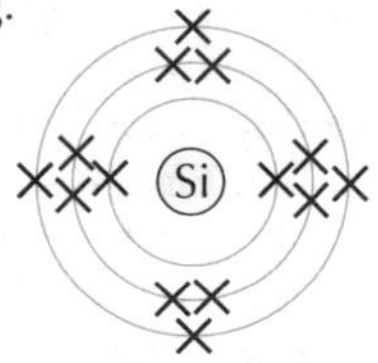

You don't need to show the electrons in exactly the same places as in the diagram above. As long as you have drawn the right number of electrons in each shell, you have the correct answer.

2 a) E.g. the inner electron shell should contain two electrons. The outer shell should only have 8 electrons.

b) 2, 8

3 a)

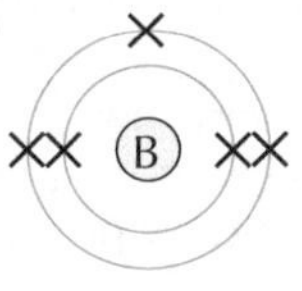

b)

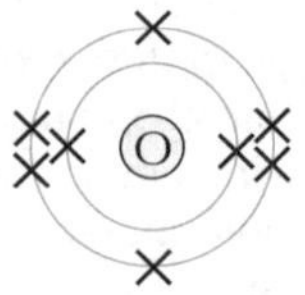

c)

d)

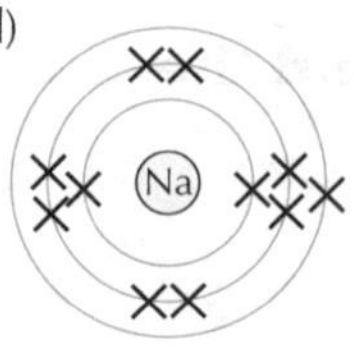

e)

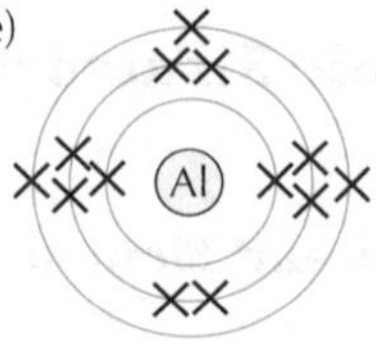

f)

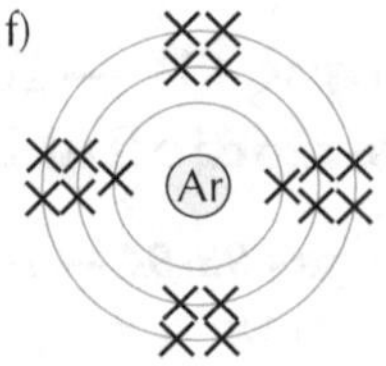

Again, what matters here is drawing the right number of electrons in each shell. Don't worry about where you've drawn them in each shell.

Pages 105-106 — The Periodic Table

Warm-Up

Early periodic tables were produced by placing the elements in order of **atomic weight**. In the modern periodic table the elements are arranged in order of **atomic number**. The columns in the table are called **groups**. These columns contain elements with **similar** chemical properties. Elements in the same column all have the same number of **electrons** in their outer shell.

B and C

1 a) radon and krypton

b) phosphorus and sodium

c) sodium

2 a)

Name	Group number	Period number	Electronic structure
Carbon	4	**2**	2, 4
Silicon	**4**	3	2, 8, 4
Nitrogen	5	**2**	2, 5
Sulfur	6	**3**	2, 8, 6
Beryllium	**2**	2	2, 2

b) i) The group number of each element is equal to the number of electrons in its outermost shell.

ii) The period number of each element is equal to the number of electron shells it has.

3 Aaliyah is correct. Elements in the same group have the same number of electrons in their outer shell which means that they react in similar ways.

Pages 107-108 — Metals and Non-Metals

Warm-Up

Metals to the left of the periodic table have **few** electrons to remove in order to be left with a full outer shell. Metals towards the bottom of the periodic table have outer electrons which are **far away from** the nucleus and so feel a **weaker** attraction to it. This means that **not much** energy is needed to remove electrons from these metals.

1 a) i) Element X

ii) Any two from: it has the lowest melting and boiling points of the three elements described. / It is brittle rather than easy to bend. / It is a poor conductor of heat.

b) Elements Y and Z.

Elements Y and Z show typical properties of metals, which form positive ions.

2 a) A, B and D

b) Element C does not conduct electricity, it has the lowest melting point of the four elements and also has the lowest density.

3 a) B and D

b) D, because it took least time for the end that wasn't near the heat source to heat up.

Pages 109-110 — Group 1 Elements

Warm-Up

It doesn't take much energy for alkali metals to lose their outer shell electron. This means they can easily form **1+** ions. Because they form ions so easily, alkali metals always form **ionic** compounds. Alkali metals react with water to produce **hydrogen** gas and a **hydroxide** solution.

1 a) i) francium
ii) lithium
b) i) B

Moving down Group 1, potassium comes between sodium and rubidium. So its melting point must be lower than sodium's melting point, but not as low as rubidium's.

ii) C

Lithium is above sodium in Group 1, and rubidium is below it. So the boiling point of sodium must be in between the boiling point of lithium and the boiling point of rubidium.

2 a) lithium + chlorine → lithium chloride
b) more vigorous

3 a) i) Metal B is the most reactive, because it takes the least time to react completely with the water and disappear.
ii) A = sodium, B = potassium, C = lithium.
b) i) sodium hydroxide and hydrogen
ii) A

Rubidium is below potassium (metal B) in Group 1. Reactivity increases moving down the group, so rubidium is more reactive than potassium. This means it takes less time for rubidium to disappear than it does for potassium to disappear.

4 Because the sodium was not stored under oil, it was exposed to the air. The sodium reacted with oxygen in the air to form a dull sodium oxide layer on its surface.

Pages 111-112 — Group 7 Elements

Warm-Up

Fluorine Chlorine Bromine Iodine Astatine

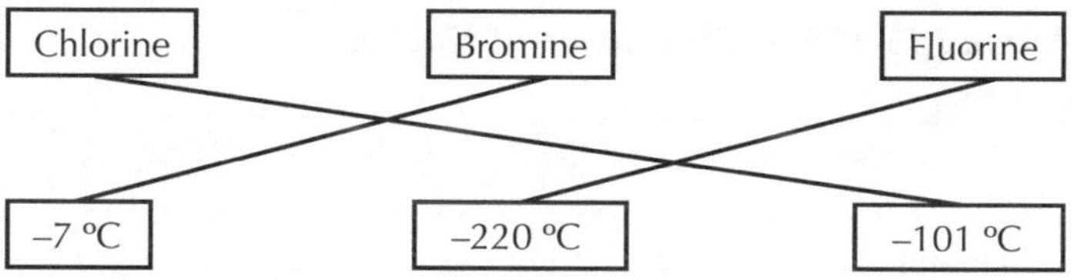

1 a) ionic bonding
b) i) slower
ii) faster

2 a) a covalent bond
b) Iodine is **below** fluorine in Group 7, and so its **outer** electron shell is further from its nucleus. This makes it harder for iodine to **gain** an electron, and so it is **less** reactive than fluorine. This makes it react more slowly and incompletely.

3 a) Bromine is more reactive than iodine so it displaces it from the potassium iodide solution. Bromine is less reactive than chlorine so it doesn't displace it from the potassium chloride solution.
b) bromine + potassium iodide → **iodine** + **potassium bromide**
c) no

Pages 113-114 — Group 0 Elements

Warm-Up

A and C

1 a) B and E could be Group 0 elements.
b) The Group 0 elements have eight electrons in their outer shell, except helium which has two.

2 D

Neon has eight electrons in its outer shell, giving it a full outer shell. This means it doesn't easily lose or gain electrons to react. So the neon gas does not react when Mariya tries to burn it.

3 a) i)

Element	Relative atomic mass	Boiling point (°C)
Helium	4	–269
Neon	20	–246
Argon	40	–186

ii) Prediction: solid
Explanation: e.g. melting point increases going down Group 0, and krypton is below argon so must have a higher melting point. So if argon is solid at –200 °C, krypton must be too.
b) Anything in the range 0.0005-0.0015 g/cm^3.

Neon comes between helium and argon as you move down Group 0, so you can predict that its density is in between their densities. The density of neon is actually 0.0009 g/cm^3.

4 Argon has a full outer shell of electrons, so will not easily give up or gain electrons, making it inert. This means that it will not react with the metal filament in the light bulb.

Topic C2 — Bonding, Structure and Properties of Matter

Pages 115-116 — Ions and Ionic Bonding

Warm-Up

Forms a positive ion	Forms a negative ion
lithium caesium magnesium	sulfur bromine fluorine

1 A

Oxygen has 6 outer-shell electrons and gains 2 so that it has a full outer shell of 8 electrons.

2 a) lithium 1
calcium 2
potassium 1
b) sulfur 2
chlorine 1
fluorine 1

3 A metal. Rhodium forms positive ions so it must be a metal.

4
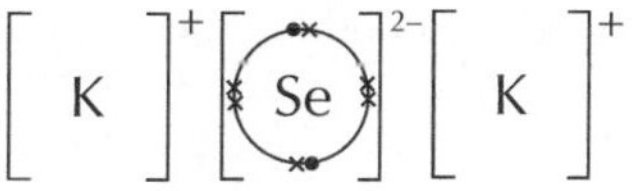

5 E.g. dot and cross diagrams don't show how the ions in caesium chloride are arranged.

6 a)
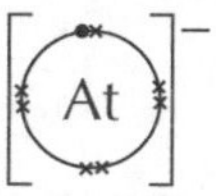

b) radon (Rn)

Pages 117-118 — Ionic Compounds

Warm-Up

C

1 a) A
b) E.g. giant ionic lattice / ionic lattice / regular lattice / lattice structure
c) i) false
ii) false
iii) true

2 BeI_2, BeS, KI, K_2S

3 a)

	When solid	When dissolved in water
Conducts electricity?	No	Yes

b) Potassium chloride is an ionic compound. When it is dissolved in water, the ions are free to move and carry an electric charge. When it is solid, the ions are held in place so it can't conduct electricity.

4 a) Iron ions: 2+, Oxide ions: 2–
b) FeO

The charges must balance. The charge of one iron ion is balanced by the charge of one oxide ion.

c) E.g. iron(II) oxide would not have a low melting point. This is because a lot of energy is needed to overcome the strong electrostatic forces between the ions / break the strong ionic bonds.

Pages 119-120 — Covalent Bonding

Warm-Up

A covalent bond forms between **two non-metal atoms**. Each atom will form enough covalent bonds to **fill its outer shell of electrons.**

1 a) C
b) Silicon must gain four electrons to get a full outer shell of eight electrons so will form four covalent bonds.

2 a) N_2H_4
b) E.g. the model shows the arrangement of atoms in space / the model shows the relative sizes of atoms.

3 D

4 a) CH_2O
b) 4

The double bond between C and O counts as two separate covalent bonds.

c)
```
H
 \
  C=O
 /
H
```

d) The dot and cross diagram/Figure 2 shows the electronic structure / which atom each of the electrons in the covalent bonds have come from.

Pages 121-122 — Simple Molecular Substances

Warm-Up

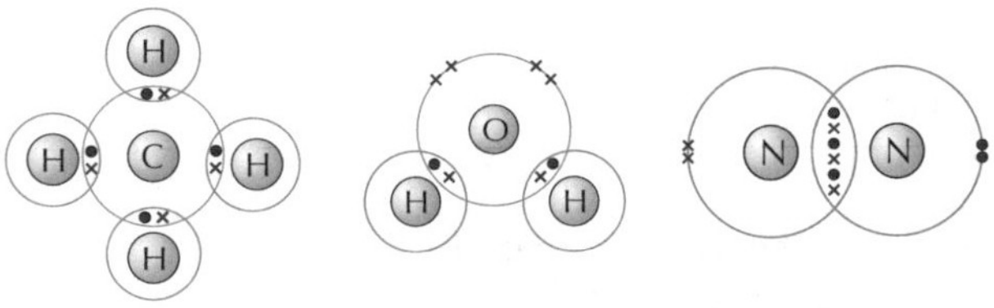

1 a) It has a high melting point / conducts electricity when liquid.
b) E.g. substance D. It has a low melting point and does not conduct electricity when liquid.

2 a)

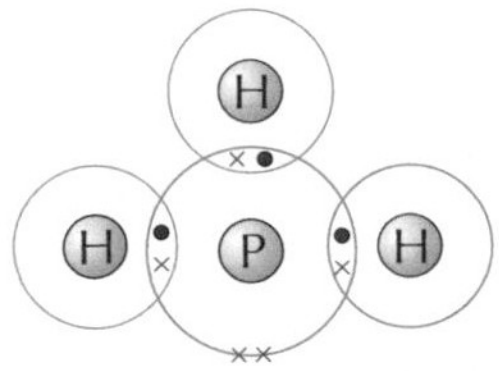

b) Diphosphane has a higher melting point than phosphane because it is a larger molecule than phosphane **cluck**.
c) B

Pages 123-124 — Polymers and Giant Covalent Structures

Warm-Up

Polymers are chains of **repeating units**. Polymers have stronger **intermolecular forces** between them than simple **covalent** molecules. This means they're usually **solid** at room temperature.

1

Name	Molecular formula of monomer	Molecular formula of polymer	Displayed formula of polymer
Poly(propene)	C_3H_6	$\mathbf{(C_3H_6)_n}$	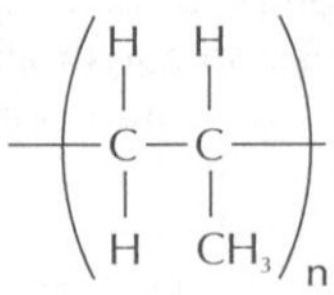
PTFE	C_2F_4	$(C_2F_4)_n$	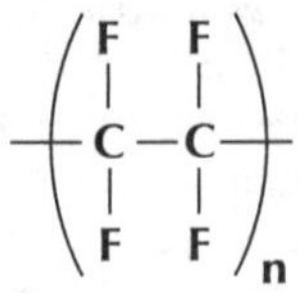
Poly(styrene)	$\mathbf{C_8H_8}$	$\mathbf{(C_8H_8)_n}$	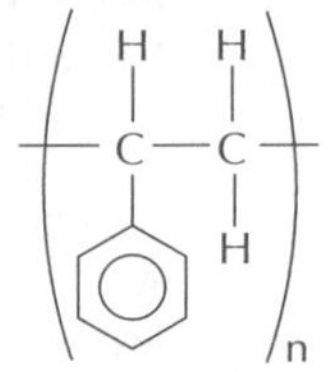

2 a) solid
b) Substance X must be made up of **two non-metals.**

3 a) i) Carbon dioxide is a **simple molecular substance.**
ii) Silicon dioxide is a **giant covalent structure.**
b) i) intermolecular forces
ii) covalent bonds
c) More energy is needed to break the covalent bonds in silicon dioxide than to overcome the intermolecular forces between carbon dioxide molecules.

Pages 125-126 — Structures of Carbon

Warm-Up

Structure	Structure	Conducts electricity?
diamond	giant covalent	no
graphite	layers of carbon atoms arranged in hexagons with no covalent bonds between layers	**yes**
graphene	single layer of carbon atoms arranged in hexagons	**yes**
fullerenes	molecules of carbon shaped like hollow balls or cylinders	cylindrical molecules conduct electricity

1 a) E.g. diamond doesn't conduct electricity.
b) E.g. graphene.

2 a) graphene
b) Structure X is strong because of the large number of covalent bonds between the carbon atoms.

3 a) i) diamond
ii) hard, does not conduct electricity, high melting point
b) i) graphite
ii) C

Pages 127-128 — Metallic Bonding

Warm-Up

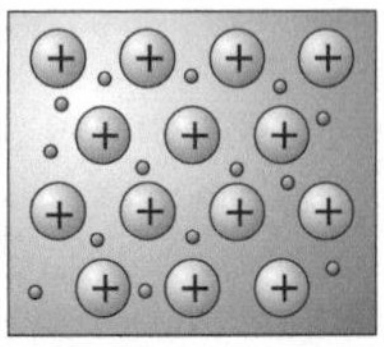

1 The layers of atoms in pure copper can easily slide over each other.

2 D

Metals are thermal conductors, so they are not good insulating materials. The layers of atoms in alloys aren't regular so they can't easily slide over each other. Electrons in metals are delocalised so are not in fixed positions.

3 a) E.g. high thermal/heat conductivity / good thermal/heat conductors.

b) Delocalised electrons in metals carry thermal/ heat energy through the structure which makes metals good conductors of heat.

4 **Solder** is harder than **lead** because it's an **alloy**. The atoms in solder are **different** sizes so the layers of atoms lose their **shape**. The metal atoms in **solder** can't slide over each other as easily as in **lead**. This means **solder** will mark **lead**.

5 a) i) high melting point

ii) There are strong forces of electrostatic attraction between the positive ions and delocalised electrons in metals. Lots of energy is needed to break apart these metallic bonds, so metals have a high melting point.

b) E.g. they are good conductors of electricity.

Pages 129-130 — States of Matter and Changing State

Warm-Up

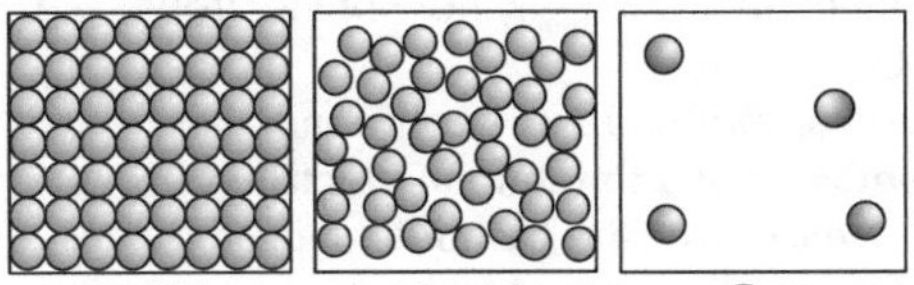

Solid Liquid Gas

1

Substance	Water	Sulfur	Mercury	Silica
Melting point (°C)	0	115	–39	1713
Boiling point (°C)	100	445	357	2950
State at room temperature	**Liquid**	**Solid**	**Liquid**	**Solid**

2 $MgCO_{3(s)} + H_2SO_{4(aq)} \rightarrow MgSO_{4(aq)} + CO_{2(g)} + H_2O_{(l)}$

3 a) glycerol

b) propane

c) i)ii)

Arrangement of particles	State	Temperature (°C)
Propanol particles are closely packed and regularly arranged.	**solid**	**–138**
Glycerol particles are spaced far apart.	**gas**	**312**
Propanol particles take up a fixed volume but don't have a fixed shape.	**liquid**	**57**
Propane particles form a fixed shape and volume.	**solid**	**–273**

d) The forces between glycerol molecules are **stronger** than the forces between propane molecules. More energy is needed to overcome the forces between **glycerol** molecules than **propane** molecules.

Topic C3 — Quantitative Chemistry

Page 131 — Relative Formula Mass

Warm-Up

Relative molecular mass = **24** + [(16 + **1**) × **2**] = 58

1 a) A_r of X = M_r of XOH – A_r of O – A_r of H
= 56 – 16 – 1 = **39**

b) Element X = potassium

Remember, you can find out the relative atomic masses of the elements using the periodic table.

2 a) M_r of NH_4NO_3 = 14 + (1 × 4) + 14 + (16 × 3)
= 14 + 4 + 14 + 48 = **80**

b) Percentage mass of an element in a compound

$= \frac{A_r \text{ of element} \times \text{number of atoms of element}}{M_r \text{ of compound}} \times 100$

So percentage mass of oxygen in ammonium nitrate

$= \frac{A_r \text{ of oxygen} \times \text{number of atoms of oxygen}}{M_r \text{ of ammonium nitrate}} \times 100$

$= \frac{16 \times 3}{80} \times 100 = \mathbf{60\%}$

Pages 132-133 — Conservation of Mass

Warm-Up

Type of atom	Number of each type of atom	
	in reactants	in products
Mg	**1**	**1**
O	**2**	**2**
H	**2**	**2**

The number of each type of atom is the same on both sides of the equation.

1 a) i) Total mass of reactants = (23 × 2) + 2 × [(1 × 2) + 16]
= 46 + (2 × 18) = **82**

ii) Total mass of products = 2 × [23 + 16 + 1] + (2 × 1)
= (2 × 40) + 2 = **82**

b) The total mass of the reactants is equal to the total mass of the products, and so mass has been conserved.

2 Mass of reactants = mass of products
Mass of iron + mass of oxygen = mass of iron oxide
So mass of oxygen = mass of iron oxide – mass of iron
= 6.3 – 4.4 = **1.9 g**

3 Mass of reactants = mass of products
Mass of lithium hydroxide + mass of nitric acid
= mass of lithium nitrate + mass of water
So mass of lithium nitrate = mass of lithium hydroxide
+ mass of nitric acid
– mass of water
= 8 + 21 – 6 = **23 g**

4 a) Experiment 2

b) In experiment 2, the total mass of products in the flask was less than the total mass of reactants added to the flask. This is because the carbon dioxide produced in the reaction escaped from the open flask and so its mass was not recorded.

Page 134 — Concentrations of Solutions

Warm-Up

concentration $= \frac{\text{mass of dissolved substance}}{\textbf{volume of solvent}}$

1 concentration $= \frac{\text{mass of dissolved substance}}{\text{volume of solvent}}$
= 2.7 ÷ 1.8 = **1.5 g/dm³**

2 concentration $= \frac{\text{mass of dissolved substance}}{\text{volume of solvent}}$
so mass of dissolved substance
= concentration × volume of solvent

volume of solvent = 250 cm³ = 250 ÷ 1000 = 0.25 dm³
so mass of copper sulfate = 32 × 0.25 = **8 g**

Topic C4 — Chemical Changes

Page 135 — Acids and Bases

Warm-Up

A

1 a) baking soda / milk

b) pH probe

2 When an acid is mixed with a base a **neutralisation** reaction takes place.
Hydrogen ions in the **acid** react with hydroxide ions in the **base**.
This reaction forms **water**.
After all of the acid and the base have reacted, the solution has a pH of **seven**.

Pages 136-137 — Reactions of Acids

Warm-Up

Metal oxides and hydroxides are bases which react with acids to form **a salt and water**.

1

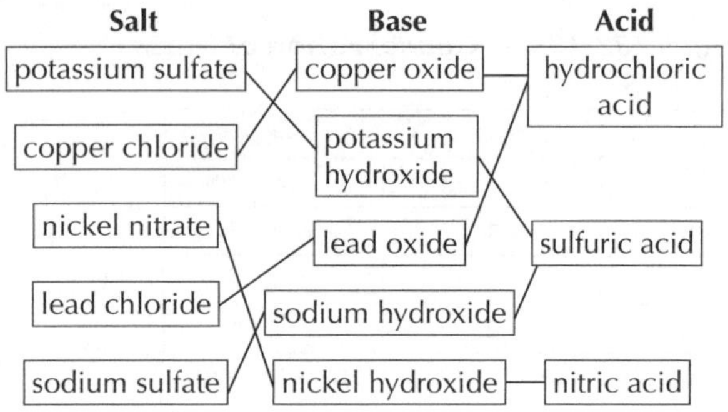

2 a) $MgO_{(s)} + \mathbf{HCl_{(aq)}} \rightarrow X_{(aq)} + H_2O_{(l)}$

b) C

c) magnesium sulfate

3 a) D

b) Filter the unreacted base from the solution of the salt. — **3**
Add insoluble base until no more reacts. — **2**
Warm dilute acid using a Bunsen burner. — **1**
Leave the solution to cool and crystallise. — **5**
Heat the salt solution over a water bath to evaporate some solution. — **4**
Filter the crystals and dry them. — **6**

4 Amir is correct. Metal carbonates react with acids to form carbon dioxide. The reaction between a metal hydroxide and an acid does not produce a gas. So Amir's salt must be copper carbonate.

Pages 138-139 — The Reactivity Series and Extracting Metals

Warm-Up

The higher up the reactivity series a metal is, the more easily it forms **a positive ion.**

hydrogen and carbon

E.g. gold / silver

1 Carbon can be used to extract metals that are **below** it in the reactivity series. Oxygen is removed from the metal oxide in a process called **reduction**. Carbon cannot be used to extract metals which are **more** reactive than it. These metals must be extracted using other methods, such as **electrolysis**.

2 A

3 a) i) aluminium

ii) Aluminium has gained oxygen forming aluminium oxide.

b) Aluminium is more reactive than carbon.

4 a) 1. bodium, 2. kandium, 3. antium

b) Kandium oxide loses oxygen.

Pages 140-141 — Reactions of Metals

Warm-Up

Metals react with acids to form **hydrogen and a salt.**

Some metals also react with water, forming a metal hydroxide and **hydrogen.**

A metal will displace a **less** reactive metal from a compound.

1 C and D

2 a) 1. sodium
2. calcium
3. iron

b) Iron doesn't react with water, so no products will have been formed.

3 a) K, M, L

b) C

The independent variable is the variable that you change. Jenson is changing the metal in each experiment to compare their reactivity, so the choice of metal is the independent variable.

c) Any two from: e.g. surface area of metal / mass of metal / starting temperature of reaction / concentration of acid / volume of acid.

d) No. M is less reactive than K so M cannot displace K from a compound.

Pages 142-143 — Electrolysis

Warm-Up

negative electrode — B, oxygen gas — D,
oxide ion, O^{2-} — E, positive electrode — F, flow of electrons — A,
aluminium ion, Al^{3+} — C

1 In the electrolysis of molten ionic compounds, an electrical current is passed through an **electrolyte**. Positive ions move towards the **cathode**. Negative ions move towards the **anode**. At the electrodes, these ions form elements. These ions are said to be **discharged** from the molten ionic compound.

2 a) The ions are free to move in molten lithium chloride, so it can conduct electricity.

b) i) chlorine

ii) lithium

3 a) Ions aren't free to move in an ionic solid. So the solid aluminium oxide is melted to allow the ions to move freely and conduct electricity.

b) Add cryolite.

c) The positive electrode/anode. It is made from carbon which reacts with oxygen forming carbon dioxide.

4 a) Br^-, OH^-, K^+, H^+

b) i) A

ii) Halide/bromide ions are present, so bromine is formed.

c) i) hydrogen

ii) Potassium is more reactive than hydrogen. This means hydrogen gas forms.

Topic C5 — Energy Changes

Page 144 — Exothermic and Endothermic Reactions

Warm–Up

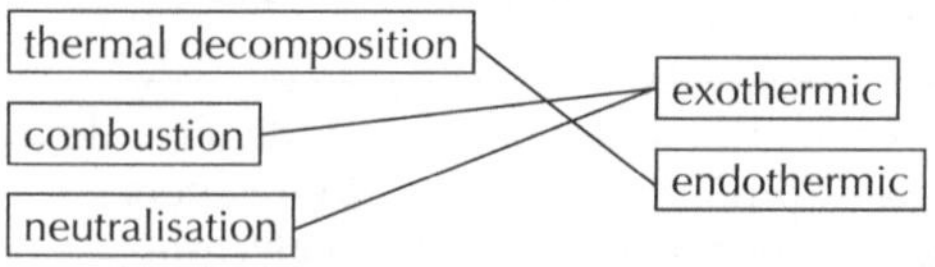

1 a) Kezia

Neutralisation reactions are exothermic so they give energy out to the surroundings.

b) Lorna has insulated her polystyrene cup / has put a lid on the cup and surrounded it with cotton wool. This means less energy is lost to the surroundings.

Page 145 — Reaction Profiles

Warm–Up

activation energy

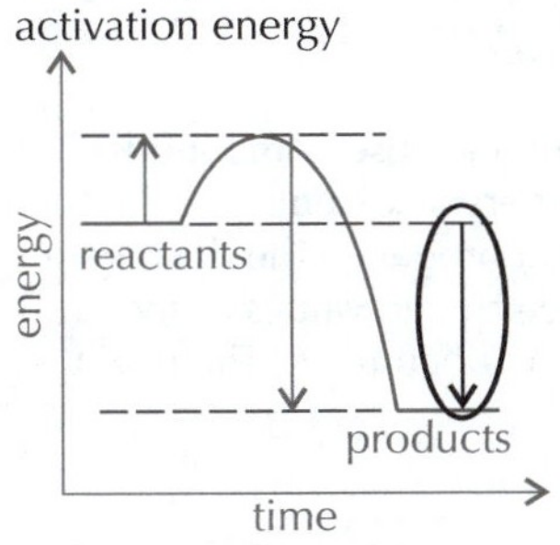

1 a) A, C, and D
b) B and E
c) B
d) C

Topic C6 — The Rate and Extent of Chemical Change

Pages 146-147 — Rates of Reaction

Warm-Up

The **higher** the temperature, the faster the rate of reaction.

A **lower** concentration or pressure will reduce the rate of reaction.

A smaller surface area of solid reactants **decreases** the rate of reaction.

A catalyst **does** change the rate of reaction.

1 a) B

b) When the temperature is increased, the particles all move **more quickly**.
This means that they will collide **more frequently**.
The particles will also have **more energy**.

c) No, the reaction equation would be the same.

Catalysts are not used up during reactions, so the reaction equation isn't affected when a catalyst is used.

2 a)

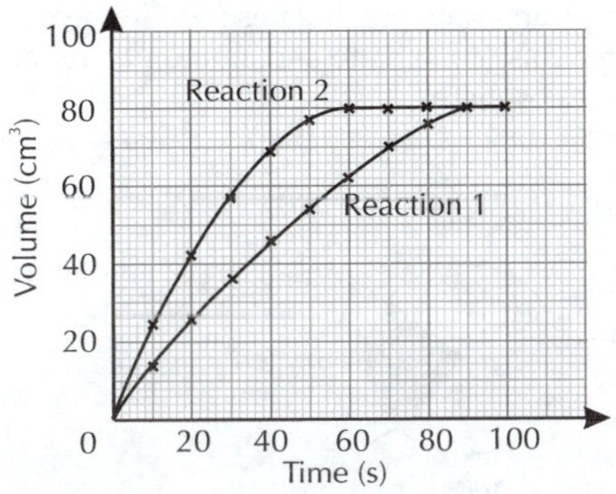

b) Reaction 2

c) E.g. the concentration of the acid / the size of the marble chips / the presence of a catalyst.

3 The fire will burn more quickly with smaller chunks of wood because they have a **bigger** surface area than a large piece of wood. This means that **more** oxygen molecules can collide with the **surface** of the wood, increasing the chance of successful collisions between reacting particles.

Pages 148-149 — Measuring Rates of Reaction

Warm-Up

B

1 a) C

b) mean rate of reaction = change in mass ÷ time taken
= 0.06 ÷ 15.5 = **0.004 g/s**

2 a) 31, in the fourth column of the table, should be circled.

b) (63 + 65 + 65) ÷ 3 = 64.333...
= **64.3 cm³** (to 1 d.p.)

c) 2.0 mol/dm³ should be circled.

d) E.g. using the wrong acid concentration / incorrectly measuring the time / incorrectly measuring the reactants used / carrying out the reaction at the wrong temperature / using a piece of calcium with a different surface area.

3 a) A

b) time taken for cross to disappear

Page 150 — Reversible Reactions

Warm-Up

A, B and D are all false.

1 a) i) B

ii) The reaction will move in the backward direction.

b) The energy transferred in the forward reaction is equal to the energy transferred in the backward reaction.

c) C

A closed system is necessary for a reversible reaction to reach equilibrium.

Topic C7 — Organic Chemistry

Pages 151-152 — Hydrocarbons and Fractional Distillation

Warm-Up

Crude oil is formed mainly of ancient **plankton** that was buried in mud.

Crude oil is a **mixture** of different molecules.

Most of the compounds in crude oil are **hydrocarbon** molecules.

Crude oil is a **finite** resource.

1 a)

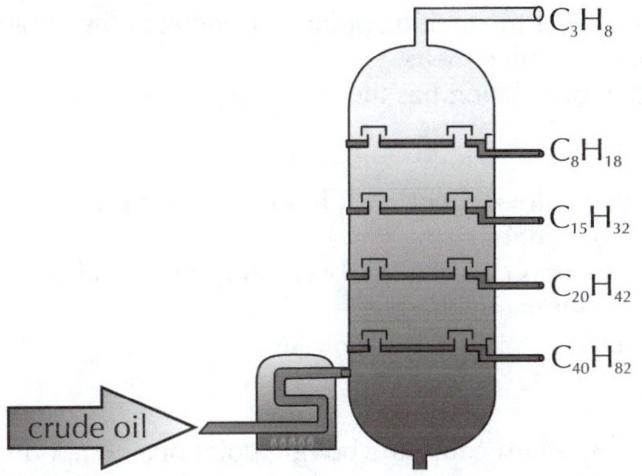

b) B

2

Name	Chemical formula	Structure
ethane	$\mathbf{C_2H_6}$	H H H–C–C–H H H
propane	C_3H_8	H H H H–C–C–C–H H H H
butane	$\mathbf{C_4H_{10}}$	H H H H H–C–C–C–C–H H H H H

3 a) $C_7H_{16} + 11O_2 \rightarrow 7CO_2 + 8\mathbf{H_2O}$

b) Heptane has a lower boiling point because it has a shorter carbon chain than triacontane.

c) Heptane is less viscous than triacontane because heptane has a shorter carbon chain.

d) triacontane

Pages 153-154 — Uses and Cracking of Crude Oil

Warm-Up

There is a higher demand for **short-chain fractions of crude oil.**
Cracking involves heating hydrocarbons to **break long-chain hydrocarbons into shorter ones.**

1 a) $C_8H_{18} \rightarrow C_6H_{14} + C_2H_4$

b) $C_{10}H_{22} \rightarrow C_6H_{14} + \mathbf{2}C_2H_4$

2 Alkanes from **crude oil** can be used to make solvents and lubricants. **Long** chain alkanes can be cracked to produce alkanes with **shorter** chains. These make better fuels. Cracking also produces **alkenes**. These can be used as the starting material to make **polymers**.

3 a) It acts as a catalyst.

b) Alkenes produced by cracking react with bromine so the bromine water turns colourless.

4 a) The hydrocarbons found in petrol are **shorter** than those in bitumen. This means they are more useful as fuels. **More** bitumen is made than is needed. Cracking bitumen **increases** the **supply of** hydrocarbons found in petrol. This means less bitumen is wasted.

b) kerosene

The supply of kerosene is higher than the demand for it, so unused kerosene can be cracked to make petrol.

Topic C8 — Chemical Analysis

Pages 155-156 — Purity and Formulations

Warm-Up

Impurities lower the melting point, and increase the range over which the substance melts.
So that the formulation has the required properties for its specific purpose.

1 a) B, D

b) Lip balm C, because it has the highest proportion of coconut oil.

2 shampoo, steel alloy, brass alloy, bronze alloy, green paint, cherryade

3 a) i) D

ii) D A B C

b) i) C

ii) Impurities cause the boiling point of a compound to increase, so the purest sample will have the lowest boiling point.

4 a) E.g. sample X doesn't melt over a range of temperatures / it has a sharp melting point, which suggests it is pure. Sample Y melts over a range of temperatures, which suggests it is less pure than sample X / impure.

b) Sample Y

Pages 157-158 — Paper Chromatography

Warm-Up

The mobile phase and the stationary phase.

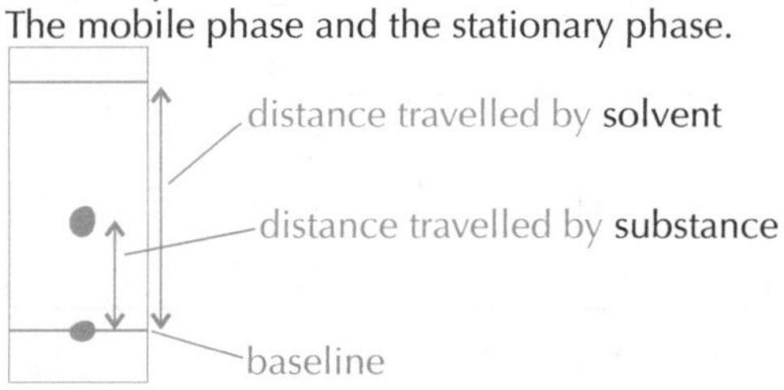

1 a) $R_f = \frac{\text{distance travelled by substance}}{\text{distance travelled by solvent}}$

R_f value of A = 31 ÷ 71 = 0.436... = **0.44** (2 s.f.)

R_f value of B = 54 ÷ 71 = 0.760... = **0.76** (2 s.f.)

b) Dye B

c) No, the food colouring produced two spots on the chromatogram which shows it contains more than one substance.

2 a) i) Distance travelled by spot E = 2.4 cm
Distance travelled by solvent = 4.0 cm

ii) $R_f = \frac{\text{distance travelled by substance}}{\text{distance travelled by solvent}}$
= 2.4 ÷ 4.0 = **0.6**

b) The R_f value would be different, because compounds have different solubilities in different solvents.

c) E.g. the ink from the printer that produced the document will form spots which have the same R_f values as the spots formed by the ink from the document. There will be no spots that don't match.

Page 159 — Tests for Gases

Warm-Up

Bubble the gas through limewater / shake the gas with limewater.
If it is carbon dioxide, the limewater will turn cloudy.

1 a) The result tells her the gas is not oxygen.

b) i) E.g. chlorine gas.

ii) If chlorine gas is present the paper will be bleached white.

c) Hydrogen

d) E.g. she should have carried out the tests in a fume hood, so that any harmful gases could not escape into the room.

Topic C9 — Chemistry of the Atmosphere

Pages 160-161 — The Evolution of the Atmosphere

Warm-Up

In the first billion years of Earth's life, gases in the atmosphere were given out by **volcanoes**.
At this time, one of the most abundant gases in the atmosphere was **carbon dioxide**.
This gas was then absorbed by green plants and algae.
These organisms produced **oxygen**.

1 In order from top of the arrow (the present) to bottom of the arrow (4.6 billion years ago): Animals begin to evolve; Carbonate precipitates form in the ocean; Small amounts of methane and ammonia are released into the atmosphere.

Answers must be in the correct order but they don't need to be correctly placed on the scale.

2 a)

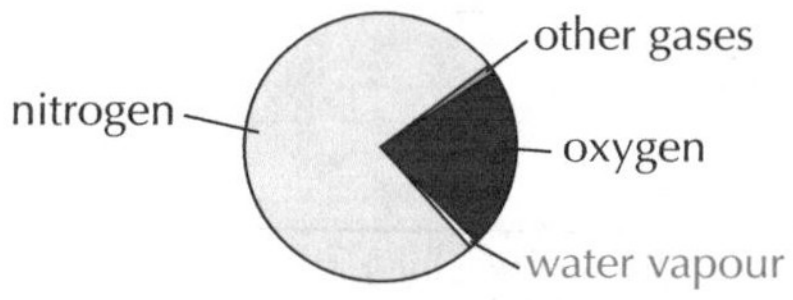

Today

b) The proportion of water vapour has decreased. This is because the water vapour condensed to form the oceans.

3 a) Organisms in the sea fall to the seabed when they die. They get squashed down by other dying organisms over **millions** of years. This process forms **sedimentary** rocks, such as limestone. Limestone is formed from the shells of dead **marine organisms** which settle on the seabed. Natural gas and **crude oil** are formed from the remains of organisms called **plankton**.

b) Coal is formed from thick layers of plants that died and settled on the seabed. So coal can't have been produced on a planet that never supported life.

Pages 162-163 — Greenhouse Gases and Climate Change

Warm-Up

Greenhouse gases keep the Earth warm enough to support life.

Any two from e.g. rising sea levels / certain areas getting too much or too little rain / more frequent and severe storms / changes in temperature and rainfall affecting food production.

The Earth's climate is complex and there are many variables.

1 a) heat detector B

b) Carbon dioxide is a greenhouse gas whereas nitrogen is not. This means that nitrogen will absorb less heat than carbon dioxide and so more heat will be detected by heat detector B.

2 a) The percentage of carbon dioxide in the atmosphere is increasing faster and faster / at an increasing rate.

b) E.g. burning of fossil fuels / agriculture / deforestation / waste creation.

c) B

Pages 164-165 — Carbon Footprints and Air Pollution

Warm-Up

False

Carbon monoxide is a product of incomplete combustion.

Nitrogen dioxide can cause acid rain.

Global dimming is an effect caused by **carbon particulates**.

1 a) E.g. harms economic development in Brazil / people like pineapples / affects jobs involved in transport of pineapples from Brazil.

b) Any three from: e.g. only fill kettle with amount of water needed for one cup of tea / walk to travel agents / change grass skirt for something warmer / turn off lights while away from house / go on holiday somewhere closer to home.

2 Carbon particulates are a product of incomplete combustion. Incomplete combustion occurs when there is a lack of available oxygen, such as in an engine.

3 E.g. sulfur

4 a) i) sulfur dioxide

ii) Fuels can contain sulfur impurities. When these fuels are burned they release sulfur dioxide.

b) B

c) When oxides of nitrogen mix with water in clouds, they form nitric acid. This causes acid rain.

Topic C10 — Using Resources

Pages 166-167 — Finite and Renewable Resources

Warm-Up

Finite resources	Renewable resources
aluminium uranium petrol	food fresh water rubber

1 B

2 Diesel comes from crude oil, which is a finite resource. This means diesel is also a finite resource and will run out too.

3 a) i) 5.0×10^5 MJ/m^3 = **500 000** MJ/m^3

ii) 5.0×10^2 MJ/m^3 = **500** MJ/m^3

iii) 500 000 – 500 = **499 500** MJ/m^3

b) Nababa Fruit skins are most likely to be considered a renewable fuel because they reform at a faster rate than they will be used at.

Page 168 — Reuse and Recycling

Warm-Up

E.g. it reduces the amount of energy used to make new glass. / It conserves fossil fuel resources which provide energy to make new glass.

Glass bottles are crushed and melted to be reshaped into new products.

1 a) 4 × 1000 = **4000 kg**

b) i) Any two from: e.g. it requires lots of energy which comes from fossil fuels. / It damages the landscape. / It produces lots of waste. / It destroys habitats.

ii) Any two from: e.g. more aluminium needs to be extracted which uses up finite resources of aluminium faster. / The amount of waste sent to landfill increases. / There is an increased need to burn fossil fuels to provide energy for mining and extracting aluminium.

Pages 169-170 — Life Cycle Assessments

Warm-Up

Companies have to **assess** the impact their processes and products will have on the **environment**. They can use this information to choose a **process** that does minimal harm. It also helps them to choose the best **materials** for the job. They have to look at the impact of each **stage** of the product's life.

1 A computer being powered by electricity.
— Using the product.
Poly(ethene) being made from ethene.
— Manufacturing the material.
A lot of plastic bottles being thrown away.
— Disposing of the product.
Oil being drilled out of the ground.
— Extracting the raw materials.
Window frames being made from PVC.
— Manufacturing the product.

2 C

3 a) Any two from: e.g. paper straws are made from a renewable resource (timber), whereas plastic straws are made from a finite resource (crude oil). / Paper straws can be recycled. / Paper straws will break down quickly if they end up in landfill.

b) A

Pages 171-172 — Treating Water

Warm-Up

Water that is safe for humans to drink.

B

1 a) Adding universal indicator to the whole sample would contaminate it, making it harder to analyse and identify the three samples.

b) Sample A is the sea water. Sea water contains more dissolved salts than rain water and tap water and the mass of crystals left in the flask was greatest for Sample A.

2 a) Desalination uses a lot of energy. This means it is very expensive and not practical for producing large amounts of fresh water. Grizeton is located near both lakes and ground water which are a more suitable way of obtaining fresh water.

b) i) The groundwater would be the best source of fresh water.

ii) The groundwater can provide a steady supply of water all year round, whereas the lakes dry up in summer.

c) E.g. water from Conisbeck lake could be passed through a wire mesh to remove large objects. It could then be passed through a filter bed to remove any smaller solids. The water could then be sterilised by bubbling chlorine gas / using UV light / using ozone gas.

d) C

Topic P1 — Energy

Pages 173-174 — Energy Stores, Systems and Conservation of Energy

Warm-Up
mechanical — the energy transfer when a force does work on an object
electrical — the energy transfer when a moving charge does work
by heating — when energy is transferred from a hotter object to a colder object

1

Situation	Energy is transferred to the...
An apple falling from a tree.	**kinetic** energy store.
A pan of soup being heated.	**thermal** energy store.
A hair tie being stretched.	**elastic potential** energy store.
A battery being charged.	**chemical** energy store.
A ball travelling upwards.	**gravitational potential** energy store.

2 C
Energy is transferred electrically by the electric current in the heater to thermal energy stores.

3 12 000 J
As the car and its surroundings are assumed to be a closed system, the energy in the car's kinetic energy store before braking = energy in the car's kinetic energy store after braking + energy transferred to the car and surroundings. So, energy in the car's kinetic energy store after braking = 41 000 − 29 000 = 12 000 J

4 a) Energy is transferred mechanically from the chemical energy stores of the weightlifter's muscles to the kinetic energy stores of her arms and the weights. Some of this energy is transferred mechanically from the kinetic energy stores of the arms and weights to the gravitational potential energy store of the arms and weights.

b) Energy is transferred mechanically from the gravitational potential energy store of the weights to their kinetic energy store.

Pages 175-176 — Kinetic and Potential Energy Stores

Warm-Up
energy in kinetic energy store = ½ × mass × (speed)² / $E_k = \frac{1}{2}mv^2$
energy in gravitational potential energy store = mass × gravitational field strength × height / $E_p = mgh$

1 a) As the cart rolls down the slope, the speed of the cart increases. This means that the amount of energy in the cart's kinetic energy store **increases**. If there are no frictional forces acting on the cart, **all** of the energy transferred out of the cart's gravitational potential energy store is transferred to the cart's kinetic energy store as it rolls down the slope.

b) $E_k = \frac{1}{2}mv^2$
$= \frac{1}{2} \times 500 \times 12^2$
$= 36\ 000$ J = **40 000 J (to 1 s.f.)**

2 a) $E_e = \frac{1}{2}ke^2$
$= \frac{1}{2} \times 280\ 000 \times 0.0050^2$
= 3.5 J

b) The energy stored in the lead's elastic potential energy store increases.
You can see from $E_e = \frac{1}{2}ke^2$ that the greater the extension of an object, e, the more energy that's stored in its elastic potential energy store.

3 a) $E_p = mgh$
$= 25 \times 9.8 \times 1.2$
$= 294$ J
= 290 J (to 2 s.f.)

b) $E_k = \frac{1}{2}mv^2$, so:
$m = 2E_k \div v^2$
$= (2 \times 150\ 000) \div 10^2$
= 3000 kg

Pages 177-178 — Specific Heat Capacity

Warm-Up
Specific heat capacity is the amount of **energy** that must be transferred to 1 kg of a substance to increase its **temperature** by **1 °C**.

1 a) true, true, false
You can see from $\Delta E = mc\Delta\theta$ that a larger mass or specific heat capacity means more energy is transferred to the oil for a particular change in temperature. The colour of the oil has no effect on the amount of energy transferred.

b) $\Delta E = mc\Delta\theta$
$= 0.025 \times 1670 \times 44$
$= 1837$ J
= 1800 J (to 2 s.f.)

c) 44 °C
The change in energy, the mass and the specific heat capacity are all the same values as those in b). So, the change in temperature must be the same as well.

2 a) Temperature at 400 J = 37.5 °C
Temperature at 0 J = 20 °C
Temperature change = 37.5 – 20 = **17.5 °C**
Temperature at 400 J can be accepted between 37 °C and 38 °C, so the temperature change can be accepted between 17 °C and 18 °C.

b) $\Delta E = mc\Delta\theta$ so:
$c = \Delta E \div m\Delta\theta$
$= 400 \div (0.018 \times 17.5)$
$= 1269.8...$
= 1300 J/kg°C (to 2 s.f.)

c) lower
The change in temperature for liquid B is greater than for liquid A, so it must need less energy per kg to change its temperature by 1 °C than liquid A.

3 $\Delta E = mc\Delta\theta$, so
$m = \Delta E \div c\Delta\theta$
$= 43\ 000 \div (1700 \times 50)$
$= 0.50588...$
= 0.51 kg (to 2 s.f.)

Pages 179-180 — Power and Reducing Energy Transfers

Warm-Up
When a system changes, some energy is **always** dissipated. This wasted energy ends up in **useless** energy stores.
1 joule of energy transferred per second

1 a) B
Room B has the thinnest walls with the highest thermal conductivity, so room B has the highest rate of cooling.

b) A
The thicker the walls and the lower the thermal conductivity, the lower the rate of cooling.

2 a) kettle B

b) Each kettle needs to transfer the same amount of energy to heat the water. Kettle B has the highest power, so it will transfer the most energy each second. This means it will transfer the energy needed in the least time.

3 a) Power of engine, $P = E \div t$
$= 12\ 500 \div 25$
= 500 W

b) Power of new engine, $P = 760$ W
$P = E \div t$, so:
$E = P \times t$
$= 760 \times 25$
= 19 000 J

c) The lubricant reduces the friction between moving parts of the moped. Friction between the parts will cause energy to be transferred to useless thermal energy stores. So, by applying a lubricant, these useless energy transfers can be reduced.

Pages 181-182 — Efficiency

Warm-Up

$$\text{efficiency} = \frac{\textbf{useful output energy transfer}}{\text{total input energy transfer}}$$

1 C

All the washing machines have the same input energy. The most efficient washing machine will have the highest useful output energy, which is C.

2 a) 4130 J

useful output energy transfer = total input energy transfer – wasted energy
= 5350 – 1220
= 4130 J

b) $\text{efficiency} = \frac{\text{useful output energy transfer}}{\text{total input energy transfer}}$
$= \frac{4130}{5350}$
$= 0.7719...$
= 0.772 (to 3 s.f.)

c) percentage efficiency = 0.7719... × 100
= 77.19...%
= 77.2% (to 3 s.f.)

3 a) B

Rearrange: $\textit{efficiency} = \frac{\textit{useful power output}}{\textit{total power input}}$ *to get:*
useful power output = efficiency × total power input

b) useful power output = efficiency × total power input
= 0.72 × 2500
= 1800 W

4 a) Find the power output of the solar panel at 25 °C using the graph:

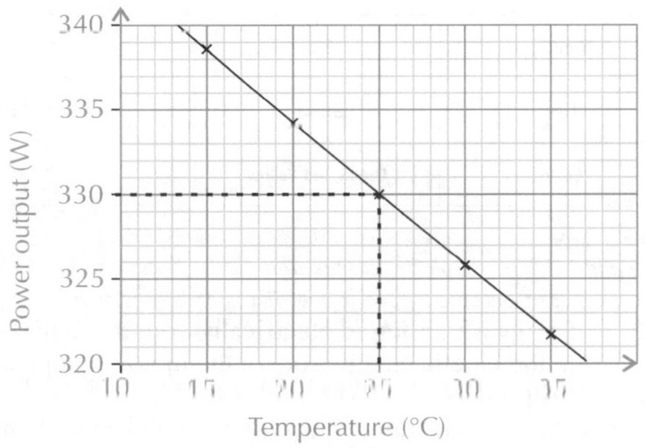

The power output is 330 W at 25 °C.
The power input is 1675 W. So the efficiency is:

$\text{efficiency} = \frac{\text{useful power output}}{\text{total power input}}$
$= \frac{330}{1675}$
$= 0.1970...$
= 0.20 (to 2 s.f.)

b) i) lower temperatures

ii) The power output is higher at a lower temperature. The power input doesn't change. You can see from $\text{efficiency} = \frac{\text{useful power output}}{\text{total power input}}$ that the higher the useful power output, the higher the efficiency. So that means the efficiency is higher at a lower temperature.

Pages 183-184 — Energy Resources and their Uses

Warm-Up

Fossil fuels: coal, natural gas, oil
Renewable resources: the tides, the Sun, water waves

1 D

Tidal barrages and wave-powered turbines would not be suitable, since the town is far away from the coast. Solar panels aren't suitable, since the weather is usually cloudy, so they wouldn't produce a reliable source of electricity. Wind turbines are most suitable, since the town regularly experiences strong winds.

2 Ben, Diego, Esme

Annie is wrong because bio-fuels do release carbon dioxide when they are burnt. Charles is wrong because coal doesn't produce nuclear waste.

3 a) i) E.g. I **agree** with Alisha because once the hydroelectric dam is constructed, it will be able to generate electricity without generating any harmful emissions/damaging the environment.

You could also have disagreed with Alisha, since the flooding caused by a hydroelectric dam causes plants to rot. When plants rot, they release greenhouse gases, leading to global warming / pollution.

ii) E.g. I **agree** with Greg because once the hydroelectric dam is built, a large area will be flooded, which will cause huge changes to the local countryside and ecosystem.

You could also have disagreed with Greg by pointing out the new ecosystems and habitats that will be formed by the construction of the dam.

iii) E.g. I **agree** with Eliza because hydroelectric power is very reliable, and its output can be quickly adjusted to meet demand.

You could also have disagreed with Eliza by saying that the local area might not get regular rainfall, and so it might not be a reliable source of electricity.

b) E.g. the energy generated by solar panels depends on the weather, so solar panels wouldn't always reliably generate enough energy for the community's needs.

You could also talk about how the energy generated may vary with the time of the day or time of the year, or that solar panels can't respond to an increase in demand.

Page 185 — Trends in Energy Resource Use

Warm-Up

Electricity use in the UK has **decreased** recently, as people are more aware about the **environmental** effects of generating electricity from fossil fuels.

1 a) E.g. because more people want to generate energy using a renewable resource instead of ones which will run out / because more people want to use energy resources that cause less damage to the environment.

b) i) B

ii) There was a large, sudden increase in the power produced by solar panels in 2006. This suggests a lot of people installed solar panels due to the government paying them.

Topic P2 — Electricity

Page 186 — Current and Circuit Symbols

Warm-Up

diode 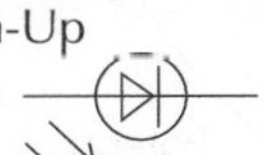

LDR 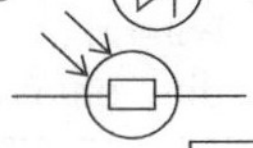

thermistor 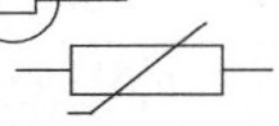

1 a) $Q = It$
$= 12 \times 5400$
$= 64\,800$ C
= 65 000 C (to 2 s.f.)

b) $Q = It$
$t = Q \div I$
$= 64\,800 \div 3.0$
$= 21\,600$ s
= 22 000 s (to 2 s.f.)

Pages 187-188 — Resistance and V = IR

Warm-Up
decreases

1 a) i) $I = V \div R$
From rearranging $V = IR$.
ii) $I = V \div R$
$= 6.0 \div 2.5$
= 2.4 A

b) $V = IR$
$R = V \div I$
$= 6.0 \div 0.75$
= 8.0 Ω

c) C
The current through resistor B has decreased (from 0.75 A to 0.5 A). Resistor B has a constant resistance. As $V = IR$, the potential difference across resistor B must have decreased.

2 a) resistance per cm = resistance of wire ÷ length of wire
= the gradient of the graph
resistance per cm = change in y ÷ change in x
= (0.30 – 0) ÷ (22.5 – 0)
= 0.0133... Ω/cm
= 0.013 Ω/cm (to 2 s.f.)
You could have used any points on the graph to calculate the gradient as they should all give you the same answer. However, try to use the largest values available as they will usually give you the most accurate answer.

b) resistance of 60 cm = resistance per cm × length of wire
= 0.0133... × 60
= 0.8 Ω

Pages 189-190 — I-V Characteristics and Circuit Devices

Warm-Up
LDRs are components whose resistance depends on light intensity.
As the light intensity increases, the resistance **decreases**.
LDRs are useful in things like **burglar detectors**.
A filament lamp

1 a) The resistance of a thermistor **decreases** as the temperature increases.

b) 9400 Ω , 12 500 Ω
Thermistors have a higher resistance at lower temperatures.
So when the temperature is 18 °C, the resistance of the thermistor must be above 9100 Ω.

2 a)

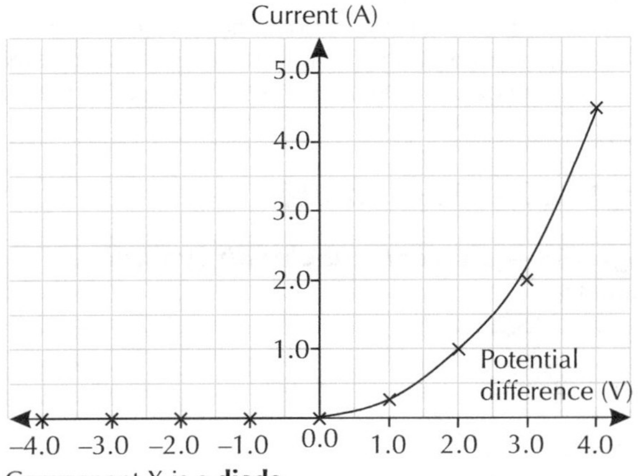

b) Component X is a **diode**.

c) Diodes only allow current to flow in one direction. So current doesn't flow in the negative potential difference direction.

Pages 191-192 — Series and Parallel Circuits

Warm-Up

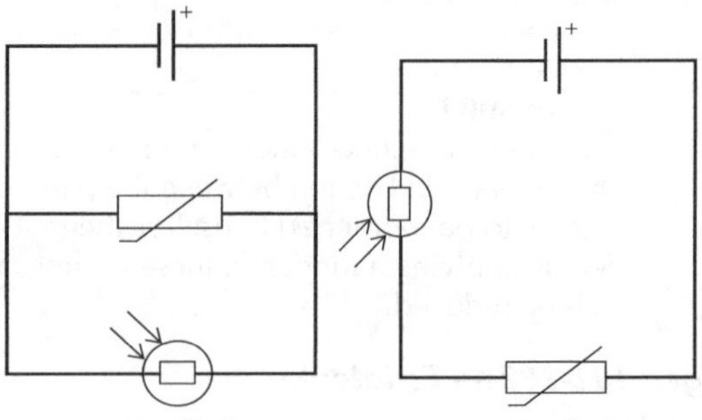

Parallel Series

1 a) 8 V
In parallel circuits, components get the full source potential difference. The battery is 8 V, so the bulb must get 8 V, so the reading on the voltmeter must also be 8 V.

b) 5 A
This is a parallel circuit, so the total current (which is 8 A through the ammeter) is equal to the sum of the currents through bulb 1 and bulb 2. Bulb 1 has 3 A flowing through it, so bulb 2 must have 5 A flowing through it.

2 a) $V_{total} = 2 + 2 + 2 = 6$ V

b) $V = IR$ so $V_{total} = I_{total} R_{total}$
$R_{total} = V_{total} \div I_{total}$
$= 6 \div 0.5$
= 12 Ω

c) 10 Ω
$R_{total} = R_1 + R_2$
$R_2 = R_{total} - R_1 = 12 - 2 = 10$ Ω

3 a) E.g. take out a bulb, and if none of the other bulbs light up when the set is connected to the mains then they are connected in series.

b) E.g. the potential difference across each bulb would be 230 V and each bulb only works if there is a potential difference of around 12 V across it.

Pages 193-194 — Electricity in the Home and Power

Warm-Up

Wire	Colour(a)	Potential difference (V)
Neutral	Blue	0
Live	Brown	(about) 230
Earth	Green and Yellow	0

1 a) A
As appliance A has a plastic casing, and plastic doesn't conduct electricity, the casing can't become live, so it doesn't require an earth wire.

b) Energy is transferred **electrically** from the mains supply to the **kinetic** energy store of the motor in appliance A.

2 a) i) Graph B
ii) The potential difference of the circuit keeps changing between positive and negative. This produces alternating current.

b) Graph B
The UK mains supply is an alternating current, so it needs an alternating potential difference in the live wire to produce this. Only graph B is showing an alternating potential difference.

3 a) $E = Pt$
$t = E \div P$
$= 900\,000 \div 1500$
= 600 s

b) Energy transferred by the Blaster in 600 s:
$E = Pt$
$= 810 \times 600$
$= 486\,000$ J
= 490 000 J (to 2 s.f.)

Pages 195-196 — More on Power

Warm-Up

$P = VI$ (power = potential difference × current)

1 a) $P = I^2R$

b) $P = I^2R$

$= 2.9^2 \times 3.8$

$= 8.41 \times 3.8$

$= 31.958$ W

= 32 W (to 2 s.f.)

c) $P = VI$

$V = P \div I$

$= 31.958 \div 2.9$

$= 11.02$ V

= 11 V (to 2 s.f.)

2

Appliance	Power (W)	Current (A)
Washing machine	**710**	3.1
Lamp	**39**	0.17
Kettle	2600	**11**
Ceiling fan	78	**0.34**

Washing machine: P = VI = 230 × 3.1 = 713 W = 710 W (to 2 s.f.)
Lamp: P = VI = 230 × 0.17 = 39.1 W = 39 W (to 2 s.f.)
Kettle: P = VI, so I = P ÷ V = 2600 ÷ 230 = 11.30... A = 11 A (to 2 s.f.)
Ceiling fan: P = VI, so I = P ÷ V = 78 ÷ 230 = 0.3391... A = 0.34 A (to 2 s.f.)

3 a) $P = VI$

$I = P \div V$

$= 50 \div 21$

$= 2.380...$ A

= 2.4 A (to 2 s.f.)

b) $P = I^2R$ so $R = P \div I^2$

$= 50 \div 2.380...^2$

$= 8.82\ \Omega$

= 8.8 Ω (to 2 s.f.)

You could also have used the potential difference given, and V = IR, to give R = V ÷ I = 21 ÷ 2.380.... = 8.82 Ω = 8.8 Ω (to 2 s.f.)

4 a) 1.43 kJ = 1430 J

$E = QV$

$Q = E \div V$

$= 1430 \div 1.5$

$= 953.333...$ C

= 950 C (to 2 s.f.)

b) E.g. The 3 V battery has a larger potential difference. $P = VI$, so power P is proportional to potential difference V. Therefore, the larger potential difference increases the torch's power and so the brightness of the torch bulb increases.

Page 197 — The National Grid

Warm-Up

Step-up transformers are used near power stations to **increase** the potential difference. The electricity is then transmitted through cables to near homes and factories where **step-down** transformers **decrease** the potential difference.

1 a) A step-down transformer.

b) This transformer **increases** the current.

2 C

Current is decreased before being transmitted through overhead cables. This reduces the energy lost during transmission, as high currents cause the cables to heat up.

Topic P3 — Particle Model of Matter

Page 198 — The Particle Model and Motion in Gases

Warm-Up

Gas — There are barely any forces between the particles so they aren't held together.

Liquid — The particles are close together but can move past each other easily.

Solid — Particles are held close together by strong forces.

1 When the temperature of the cylinders decreases, the average energy in the **kinetic** energy stores of the helium particles decreases. This means the particles move **slower** and collide with the walls of the container **less** often. So the pressure of the helium gas **decreases**.

Pages 199-200 — Density of Materials

Warm-Up

density = mass ÷ volume

1 a) C

Of the options, only the balances will measure mass. Since the toy is likely to have a mass of only a few hundred grams, a balance that only measures to the nearest kg would not be suitable. So, the balance that measures to the nearest gram would be most suitable.

b) volume of toy = volume of water after toy added – volume of water before toy added

= 400 – 300

= 100 cm³

2 a) density = mass ÷ volume

$= 6.7 \div 0.015$

$= 446.66...$ kg/m³

= 450 kg/m³ (to 2 s.f.)

b) B

The density of an object depends on what it's made of and how its particles are arranged. Since both pieces of the plank are made from the same material, they will have the same density as each other.

3 a) Rearrange density = mass ÷ volume for mass, so:

mass = density × volume

$= 1800 \times 6.5 \times 10^{-5}$

$= 0.117$ kg

= 0.12 kg (to 2 s.f.)

b) i) The hollow ball will have a lower average density than the solid ball.

ii) E.g. the air particles inside the hollow ball will be much further apart than the particles that make up the solid ball. There will therefore be (many) more particles per unit volume in the solid ball, which means it will have a higher density.

Pages 201-202 — Internal Energy, State Changes and Latent Heat

Warm-Up

the specific latent heat of vaporisation

1 C

No energy is being transferred to the kinetic energy stores of the water particles, so there is no rise in the water's temperature. So all the energy supplied by the fire is going into breaking bonds to change the state of the water.

2 The energy supplied by the kettle increases the water's **internal** energy. This causes the temperature of the water to **increase**. After the temperature gets high enough, the water begins to **boil**. The energy being transferred is now being used to **break** bonds between particles. The temperature will **stay constant** during this process.

3 a) i) 60 °C

ii) Melting

b) The mass of wax at the start was also 1 kg, because there is no change of mass during a change of state.

4 a) energy = mass × specific latent heat, rearranged gives:
mass = energy ÷ specific latent heat
= 520 000 ÷ 334 000
= 1.55... kg
= **1.6 kg (to 2 s.f.)**

b) 32 g = 0.032 kg
energy = mass × specific latent heat
= 0.032 × 334 000
= 10 688 J
= **11 000 J (to 2 s.f.)**

Topic P4 — Atomic Structure

Page 203 — Developing the Model of the Atom

Warm-Up

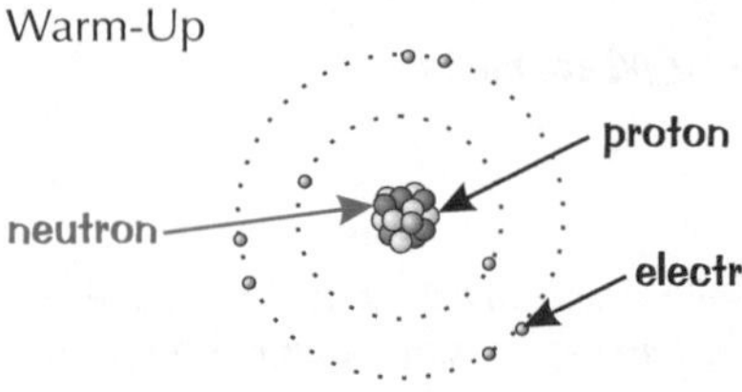

1 a) it has a positive charge.

b) E.g. the alpha particle is repelled (by the nucleus of a gold atom). So the nucleus must have a positive charge / have the same type of charge as the alpha particle.

2 The electron will have moved up to a higher energy level. This means it will be further from the nucleus and so the radius of its orbit will have increased.

Pages 204-205 — Isotopes, Radiation and Nuclear Equations

Warm-Up

Isotopes are atoms which have the same number of **protons** but different numbers of **neutrons**.

Alpha particle — 2 neutrons and 2 protons — the same as a helium nucleus.

Beta particle — An electron.

Gamma ray — A type of electromagnetic radiation.

1 a) Protons: 1, Neutrons: 2

b) Protons: 6, Neutrons: 8

c) Protons: 7, Neutrons: 7

2 Source X = gamma radiation
Source Y = alpha radiation
Source Z = beta radiation

3 a) $^{222}_{86}\text{Rn} \rightarrow ^{222}_{86}\text{Rn} + ^{0}_{0}\gamma$

The mass and charge of an atom doesn't change when it emits a gamma ray.

b) $^{234}_{90}\text{Th} \rightarrow ^{234}_{91}\text{Pa} + ^{0}_{-1}\text{e}$

4 a) Alpha particles would be absorbed by all thicknesses of metal and would not reach the detector.

b) No. Gamma radiation would pass straight through the metal so the amount reaching the detector wouldn't vary much with the thickness of the metal.

Pages 206-207 — Half-life

Warm-Up

A

1 a) B

Activity after one half-life = 8000 ÷ 2 = 4000 Bq
Activity after two half-lives = 4000 ÷ 2 = 2000 Bq
It takes 80 seconds for the activity to reach 2000 Bq, so 2 half-lives = 80 seconds. Half-life = 80 ÷ 2 = 40 s

b) Activity after two half-lives = 2000 Bq (calculated in a)).
After three half-lives activity = 2000 ÷ 2 = 1000 Bq
After four half-lives activity = 1000 ÷ 2 = 500 Bq
So it takes **four** half-lives to fall below 600 Bq.

2 From Table 1, you can see it takes 10 minutes for the count rate of the sample to halve,
so half-life = 10 minutes.
count rate at 20 mins = 1550 cpm
count rate at 30 mins = 775 cpm
count rate at 40 mins = 387.5 cpm
count rate at 50 mins = 193.75 cpm
count rate at 60 mins = 96.875 cpm
So the count rate won't be below 20 cpm after an hour.

3 a) E.g. the half-life is the time taken for the activity to halve. The initial activity of the sample is 2.4×10^3 Bq. $2.4 \times 10^3 \div 2 = 1.2 \times 10^3$. Reading from Figure 1, it takes 80 s for the activity to reach 1.2×10^3 Bq.
So half-life = **80 s**.

b) i) The initial activity is 2.4×10^3 Bq.
After 3 half-lives this will have halved three times.
After 1 half-life, activity = 1200 Bq
After 2 half-lives, activity = 600 Bq
After 3 half-lives, activity = **300 Bq**

ii) You know from a) that one half life = 80 s, and that 3 half-lives have passed.
So the time taken = 3 × 80 = **240 s**

You could have also read the time taken from the graph in Figure 1.

c) After 1 half-life, half the original unstable isotope atoms remain.
So the mass remaining = 28 ÷ 2 = 14 g.
After two half lives the number of unstable atoms will have halved again.
So mass remaining = 14 ÷ 2 = **7 g**

d) Y

Sample B has the same initial activity as sample A, so line X isn't correct. Sample B has a shorter half-life than sample A, so the activity must decrease more quickly than sample A. Only line Y out of lines X, Y and Z does this.

Pages 208-209 — Irradiation and Contamination

Warm-Up

False, False, True

Contamination is where atoms of a radioactive substance are on or inside an object.

1 a) E.g. burying the waste deep underground means there's a lot of ground between people and the waste, which could absorb radiation emitted by the waste. / Placing the waste far away from people will help stop them being exposed to radiation emitted by the waste.

b) E.g. I agree. Having the waste in sealed containers means none of the waste is able to touch other objects/people.

2 a) E.g. in a sealed, lead-lined container. The lead will absorb radiation emitted by the source, and so stop any radiation leaving the box and irradiating its surroundings. Having a sealed box means that none of the sample will be able to get out of the box and contaminate people/ objects.

b) E.g. he isn't wearing protective gloves/using tongs to handle the sample. This increases the chance of radioactive atoms getting onto his hands and his hands becoming contaminated. / He is pointing the sample directly into the face of the other scientist. This exposes the scientist to the radioactive material, which means she could easily become irradiated by it.

The other scientist is also increasing the risk of irradiation by not looking where she is pointing the source. However, she has taken better safety precautions than the careless scientist.

c) E.g. as the radioactive atoms are on his hands, there's a high risk that some of these atoms will be transferred to the sandwich and then into his body, where the ionising radiation emitted by the radioactive atoms can cause a lot of damage.

Topic P5 — Forces

Page 210 — Contact and Non-Contact Forces

Warm-Up

Vector quantities — Have magnitude and a direction.
Scalar quantities — Have magnitude but no direction.
friction

1 a) i) A
ii) A non-contact force.
b) The magnitude (size) of the forces.
c) same size

The arrow lengths for forces A and B are both the same, so the forces are the same size.

Pages 211-212 — Weight, Mass and Gravity

Warm-Up

An object's mass changes depending on where in the universe it is.

The mass of an object is always the same, no matter where it is. It's the weight of an object that will change depending on where it is in the universe.

$W = mg$ (weight = mass × gravitational field strength)

1 a) $W = mg$
$= 2.0 \times 9.8$
= 19.6 N
= 20 N (to 2 s.f.)
b) A
c) If the student did the same experiment on the moon, the object would weigh **less**.

The object's weight is directly proportional to gravitational field strength ($W = mg$) for a given mass. So the weight measured would be lower as the gravitational field strength, g, is weaker on the Moon than on Earth.

2 a) $W = mg$
$m = W \div g$
$= 540 \div 9.8$
$= 55.102...$
= 55 kg (to 2 s.f.)
b) First calculate Peri-Peri's new weight:
$W = mg = 47 \times 9.8 = 460.6$ N
change in weight = old weight – new weight
$= 540 - 460.6$
$= 79.4$
= 79 N (to 2 s.f.)

3 a)

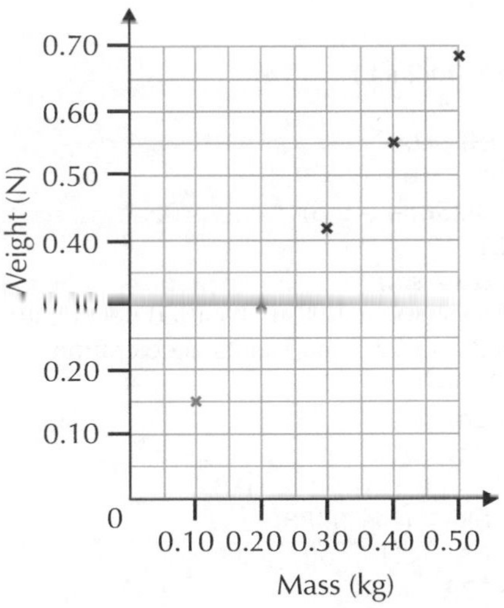

b) E.g.

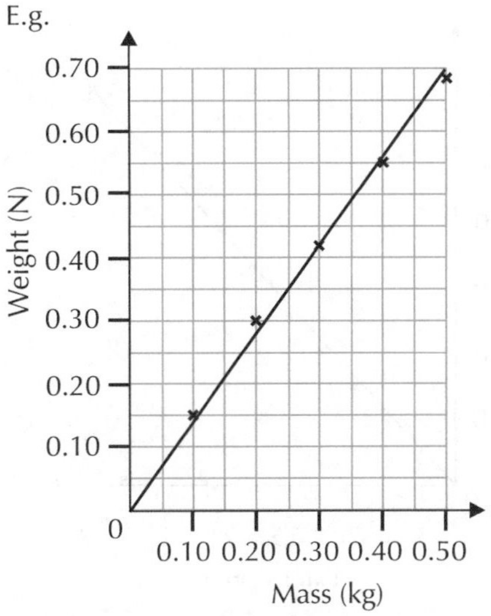

c) The gradient of the graph is equal to weight ÷ mass.
$W = mg$ so $g = W \div m$ so the gradient will give the gravitational field strength of Titan.
gradient = change in y ÷ change in x
e.g. $= 0.70 \div 0.50$
= 1.4 N/kg

Your answer will depend on your line of best fit.
Accept answer between 1.36-1.44 N/kg.

Page 213 — Resultant Forces and Work Done

Warm-Up

If a number of forces act on an object, they can be replaced by a **single** force called the **resultant** force.

1 a) resultant force = 365 N – 53 N = **312 N**
b) The work done by the dog against friction causes energy to be transferred to the **thermal** energy store of the skateboard. This causes the overall temperature of the skateboard to **increase**.
c) $W = Fs$
$s = W \div F$
$= 525 \div 38$
$= 13.815...$ m
= 14 m (to 2 s.f.)

Pages 214-215 — Forces and Elasticity

Warm-Up

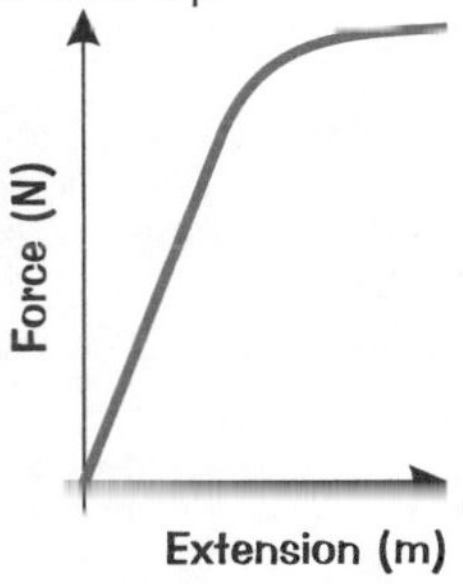

1 a) C

Force and extension are not directly proportional after a spring has been stretched past its limit of proportionality. Therefore, the relationship between force and extension is no longer linear.

b) i) $W = ke$

The weight of the fish = the force stretching the spring.
The force stretching the spring is $F = ke$

ii) $k = 1600$ N/m, $e = 0.068$ m
$W = ke$
$= 1600 \times 0.068$
$= 108.8$
= 110 N (to 2 s.f.)

2 a)

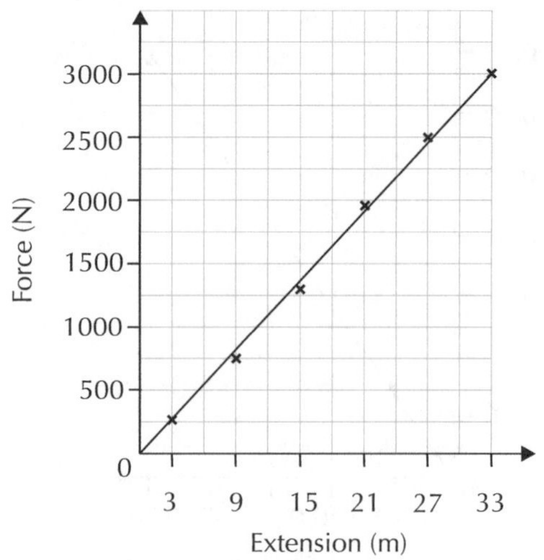

The line of best fit goes through the origin as there is zero force at zero extension. Force and extension show a directly proportional relationship, so the gradient will be equal to the spring constant.

b) The gradient of the graph is equal to force ÷ extension.
$F = ke$ so $k = F \div e$ so the gradient will give the spring constant of the bungee rope.
gradient = change in y ÷ change in x
= 3000 ÷ 33
= 90.9090...
= 90 N/m (to 1 s.f.)

c) $E_e = ½ke^2$
$e^2 = E_e \div (½ \times k)$
= 65 700 ÷ (½ × 90.9090...)
= 65 700 ÷ 45.4545...
= 1445.4
$e = \sqrt{1445.4}$
= 38.018...
= 40 m (to 1 s.f.)

Pages 216-217 — Distance, Displacement, Speed and Velocity

Warm-Up
15 m to the left, 18 cm north
Displacements contain magnitude and direction.

1 a) i) 6 m/s
ii) $s = vt$
= 6 × 300
= 1800
= 2000 m (to 1 s.f.)

b) $s = vt$
$v = s \div t$
= 230 ÷ 18
= 12.7...
= 13 m/s (to 2 s.f.)

2 D
12 km = 12 000 m
$s = vt$ so $v = s \div t$ = 12 000 ÷ 1200 = 10 m/s.
The train is going from east to west. Therefore the velocity is 10 m/s due west. A is not a velocity as it doesn't contain a direction.

3 a) The car moves 3 lines of 2 m (= 6 m) in 0.3 seconds. Therefore the time between the two photos:
$s = vt$
$t = s \div v$
= 6 ÷ 20
= 0.3 s

b) C
A is true because speed is just how fast the car is going, and the car was going at 20 m/s in a) too. B is true because $s = vt$, so $v = s \div t$ = 200 ÷ 10 = 20 m/s. C is false because velocity has both magnitude and direction, and so the velocity is different to a) because the car is travelling in the opposite direction.

c) The distance travelled by the car as it drives between points A and B is **bigger than** the displacement of the car.
Distance is just how far the car has travelled along the track between points A and B, whereas displacement is the distance in a straight line between points A and B. So the distance is bigger than the displacement.

4 27 days in seconds = 86 400 × 27 = 2 332 800 s
Average speed of the Moon:
$s = vt$
$v = s \div t$
= (2.4×10^9) ÷ 2 332 800
= 1028.806...
= 1000 m/s (to 2 s.f.)

Pages 218-219 — Acceleration

Warm-Up
Deceleration is when an object **slows down.**

1 a) $a = \Delta v \div t$

b) Typical running speed $v \sim 3$ m/s
The runner starts from rest so $u = 0$ m/s
$a = \Delta v \div t$
= (3 – 0) ÷ 4
= 0.75
= 0.8 m/s² (to 1 s.f.)
A good approximation for typical running speed is between 2 and 3.5 m/s, so any acceleration between 0.5 and 0.9 m/s² can be accepted, depending on the value chosen for typical running speed.

c) **0 m/s²**
There is no change in the runner's velocity (his speed doesn't change and his direction doesn't change) so the runner doesn't accelerate.

2 $a = \Delta v \div t$
$t = \Delta v \div a$
= (25 – 12) ÷ 3.7
= 13 ÷ 3.7
= 3.513...
= 3.5 s (to 2 s.f.)

3 a) D
The egg fell freely under gravity so it accelerated due to gravity.
So acceleration = 9.8 m/s².

b) The egg was dropped so it started from rest.
So $u = 0$ m/s.
$a = \Delta v \div t$
$\Delta v = a \times t$
$\Delta v = v - u = v - 0 = v$
$v = a \times t$
= 9.8 × 3.3
= 32.34
= 32 m/s (to 2 s.f.)

c) $v^2 - u^2 = 2as$
$s = (v^2 - u^2) \div 2a$
= $(32.34^2 - 0^2) \div (2 \times 9.8)$
= (1045.8756) ÷ (19.6)
= 53.361
= 53 m (to 2 s.f.)

4 a) The tractor's velocity before it started decelerating is u.
It decelerates at 1.92 m/s² so its acceleration:
$a = -1.92$ m/s²
$v^2 - u^2 = 2as$
$u^2 = v^2 - 2as$
= $3.22^2 - (2 \times -1.92 \times 10.82)$
= 10.3684 – (–41.5488)
= 10.3684 + 41.5488
= 51.9172
$u = \sqrt{u^2} = \sqrt{51.9172}$ = 7.2053... = **7.21 m/s (to 3 s.f.)**

b) $a = \Delta v \div t = (v - u) \div t$
$t = (v - u) \div a$
= (3.22 – 7.2053...) ÷ (–1.92)
= (–3.9853...) ÷ (–1.92)
= 2.0757...
= 2.08 s (to 3 s.f.)

Pages 220-221 — Motion Graphs and Terminal Velocity

Warm-Up

Acceleration

An object falling through a fluid first accelerates due to the force of **gravity**. The fluid opposes this force with a drag force. Eventually, the **resultant** force on the object will be zero and the object moves at a **constant** speed, called its terminal velocity.

1 A — The tram travels at a constant speed for 3 s and then decelerates for the next 3 s.
B — The tram remains stationary for 3 s and then travels at a constant speed for the next 3 s.
C — The tram travels at a constant speed for 3 s and then travels at a slower constant speed for the next 3 s.

2 True

resultant force acting on the skydiver = air resistance − weight
At terminal velocity, the resultant force of the skydiver is zero.
Therefore, air resistance = weight.

3 a) 120 s

This is the initial distance covered before she stops for 20 s (at 200 m).

b) Speed = gradient of graph (when initially walking to the playing field) = distance ÷ time = 200 ÷ 120
= 1.666... m/s = **1.7 m/s (to 2 s.f.)**

c)

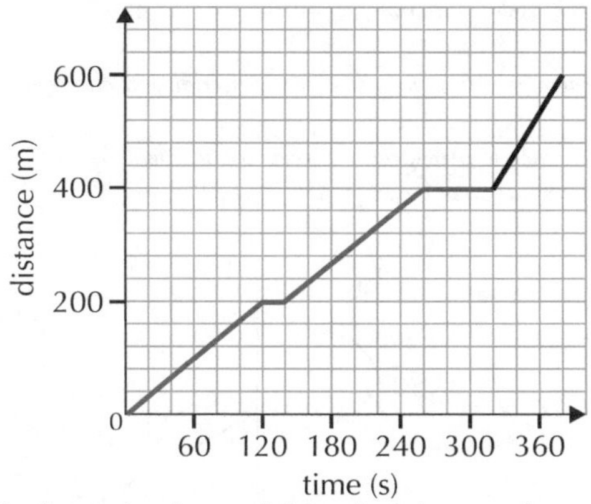

She runs back at twice the speed she walked there, so she covers the same distance (200 m) in half the time (½ × 120 s = 60 s).

Pages 222-223 — Newton's First and Second Laws

Warm-Up

The object accelerates.

F = *ma* (resultant force = mass × acceleration)

a = F ÷ m and m = F ÷ a are also correct answers.

1 a) Not balanced

The speed of the ball is changing. Therefore its velocity is changing.

b) Not balanced

The motorbike moves in a circle so it is constantly changing direction. Therefore its velocity is changing.

c) Balanced

Neither the ice skater's speed nor their direction is changing. Therefore their velocity doesn't change.

2 D

Find the acceleration of the cars:
F = ma, so a = F ÷ m
Newton 9000: a = 4160 ÷ 800 = 5.2 m/s²
Einstein 6i: a = 1100 ÷ 1560 = 0.7051... = 0.71 m/s² (to 2 s.f.)
CGP TT: a = 3090 ÷ 950 = 3.2526... = 3.3 m/s² (to 2 s.f.)
Heisenberg LM: a = 10 600 ÷ 2000 = 5.3 m/s²
So the Heisenberg LM has the greatest acceleration.

3 a) 1. The bird could change direction.
2. The bird could change speed / speed up / slow down.

There is a resultant force acting on the bird so, according to Newton's First Law, its velocity must change. Velocity is both speed and direction.

b) Resultant force *F* = –0.057 N
mass of the bird = 20.0 g = 20.0 ÷ 1000 = 0.020 kg
F = *ma* so re-arranging Newton's Second Law:
a = *F* ÷ *m* = –0.057 ÷ 0.020 = –2.85 m/s²
= **–2.9 m/s² (to 2 s.f.)**

Pages 224-225 — Newton's Third Law and Investigating Motion

1 The size of the force exerted by the man on the wall is **the same as** the size of the force exerted by the wall on the man.

2 B

Aria's hands exert a normal contact force on the barbell so by Newton's Third Law, an equal and opposite normal contact force must be exerted by the barbell on Aria's hands.

3 a) –4.2 N

The force will be equal (4.2 N) and opposite (negative sign).

b) When the ball is lying on the table, the Earth exerts a gravitational force on the ball, and the ball exerts a **gravitational** force on the Earth. These two forces are **equal** in size and act in **opposite directions.**

4 a)

X (°)	*F* (N)	*u* (m/s)	*v* (m/s)	*t* (s)	*a* (m/s²)
30	0.36	0.17	0.67	1.00	0.50
35	0.41	0.23	0.69	0.81	**0.57**
40	0.46	0.26	0.75	0.77	0.64
45	0.51	0.34	0.81	0.66	**0.71**

a = Δv ÷ t = (v − u) ÷ t
For 35° : (0.69 − 0.23) ÷ 0.81 = 0.5679... = 0.57 m/s² (to 2 s.f.)
For 45° : (0.81 − 0.34) ÷ 0.66 = 0.7121... = 0.71 m/s² (to 2 s.f.)

b)

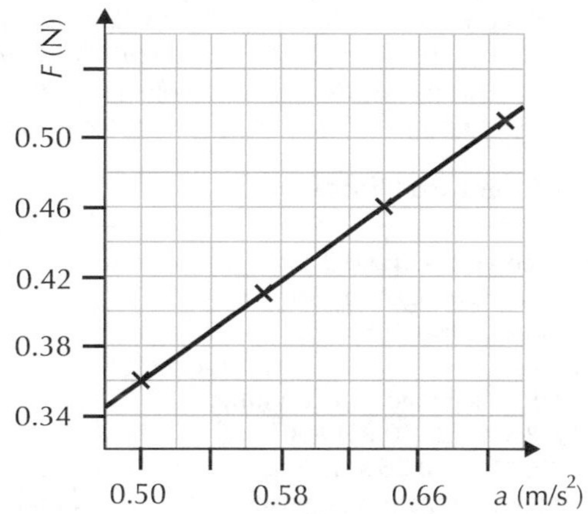

c) For the mass of the trolley, use Newton's Second Law:
F = ma
m = *F* ÷ a = gradient

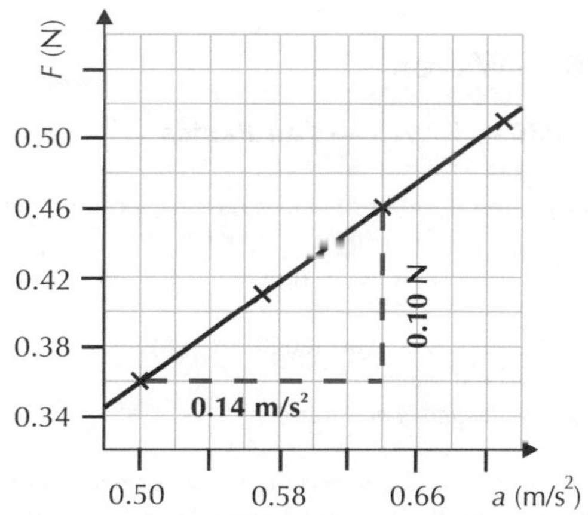

Gradient = change in *y* ÷ change in *x*
= 0.10 ÷ 0.14
= 0.714...
= **0.71 kg (to 2 s.f.)**

The mass of the trolley can be accepted between 0.71 and 0.73 kg.

Pages 226-227 — Stopping, Thinking and Braking Distances

Warm-up

Braking distance, stopping distance

1 a) 29 m

b) 45 m

Stopping distance = thinking distance + braking distance
= 16 + 29
= 45 m

c) C

A would increase the braking distance. B and D wouldn't change anything.

d) If the ostrich had been on the road in the same conditions two weeks ago then Hafsa would have been **more likely** to hit it.

2 a) When the brake pedal is pushed, the brake pads are pressed onto the wheels causing **friction**. This causes work to be done by the brakes, transferring energy from the **kinetic** energy stores of the wheels to the **thermal** energy stores of the brakes. This causes the brakes to heat up.

b) A, D

The other two are not possible outcomes of a large deceleration.

3 E.g. a driver's reaction time is the same at both speeds. However, the thinking distance of a vehicle depends on a driver's reaction time and the vehicle's speed. The thinking distance is shorter at lower speeds. Therefore, a vehicle will be more likely to stop before a collision at 30 mph compared to 40 mph, so the village will be safer.

You could also talk about the braking distance, as this is also shorter at lower speeds. This means that a vehicle is more likely to stop before a collision at 30 mph rather than 40 mph, which makes the village safer.

Page 228 — Reaction Times

Warm-up

Any value between 0.2 s and 0.9 s.

1 a) Person A's average reaction time:
$= (0.80 + 0.76 + 0.69) \div 3$
$=$ **0.75 s**
Person B's average reaction time:
$= (0.32 + 0.22 + 0.25) \div 3$
$= 0.2633... =$ **0.26 s (to 2 s.f.)**

b) Scientist C's reaction times increased after drinking alcohol. This shows that drinking alcohol can increase a driver's reaction time, which would increase the thinking distance of the driver and their vehicle. This suggests it is less safe to drive after drinking alcohol.

Topic P6 — Waves

Pages 229-230 — Waves — The Basics

Warm-Up

The **frequency** is the number of waves passing a point each second.

The time it takes for a wave to complete a full cycle is called the **period**.

wave speed = frequency × wavelength / $v = f\lambda$

1 a) Longitudinal wave

All sound waves are longitudinal waves.

b) A and C

Amplitude = the maximum displacement of the wave from 0 mm, which is 0.4 mm. Wavelength = the distance between two peaks = 4 cm.

2 a) 3 Hz

The frequency is the number of bobs each second. The bird bobs up and down three times a second, so the frequency is 3 Hz.

b) $T = 1 \div f = 1 \div 3 = 0.333... =$ **0.3 s (to 1 s.f.)**

c) The small piece of bread is one wavelength away = λ
$v = f\lambda$, so $\lambda = v \div f$
$= 0.5 \div 3$
$= 0.1666...$
$=$ **0.2 m (to 1 s.f.)**

d) No, the bird doesn't travel with the water wave. This is because water isn't carried away when water waves travel.

When a wave travels through a medium, overall the particles stay in the same place.

3 a) $T = 1 \div f$, so $f = 1 \div T$
$= 1 \div (1.5 \times 10^{-10})$
$= 6.666... \times 10^{9}$ Hz
$=$ **6.7 × 10⁹ Hz (to 2 s.f.)**

b) $v = f\lambda$, so $\lambda = v \div f$
$= (3.0 \times 10^{8}) \div (6.666... \times 10^{9})$
$=$ **0.045 m**

Pages 231-232 — Investigating Waves and Refraction

Warm-Up

A

A is correct as the wavelength can be measured directly from the string with a ruler. B is wrong as the wave speed in the string will be constant. C is wrong as the frequency is set by the signal generator and cannot be determined just by looking at the string.

1 a) A

When light refracts, it passes into the new material and changes direction.

b) **50°**

The angle of incidence is measured between the incident ray and the normal. Accept any answer between 49° and 51°.

c)

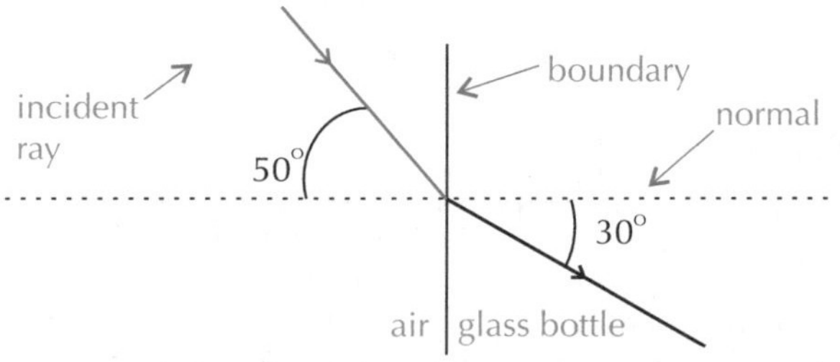

The refracted ray should be drawn below the normal and at an angle of 30° to the normal.

2 a) i) The distance between each crest is the same and is equal to the wavelength. There are two wavelengths between A and B.
$\lambda = 16 \div 2 =$ **8 cm**

ii) Frequency = the number of waves passing a point each second. Every second, two waves reach the edge of the ripple tank, so frequency = 2 Hz.
$\lambda = 8$ cm $= 0.08$ m
$v = f\lambda = 2 \times 0.08 = 0.16 =$ **0.2 m/s (to 1 s.f.)**

b) If the student increases the number of drops they drop onto the water every second, the frequency of the water waves **increases**. The speed of the water waves will remain the same, and their wavelength decreases. If a photo was taken in this case, there would be **more** lines on the photo than on the one in Figure 2.